WORLD PROBLEMS

A Topic Geography

D1809485

NAME	Form	Year
M. Edwards.	5alpha	
M. Gravey	10m₀	83/84

SECONDARY SCHOOL SERIES

General Editors: **Doris M. Lee, M.A., Ph.D.**
Professor of Education
University of London Institute of Education

J. F. Eggleston, B.Sc.
Leicester University School of Education

WORLD PROBLEMS
A Topic Geography

M. LONG, M.A.
University of London Institute of Education

B. S. ROBERSON, M.A., B.Sc. (Econ.)
University of London Institute of Education

HODDER AND STOUGHTON
LONDON SYDNEY AUCKLAND TORONTO

ISBN 0 340 22177 1

First published 1969
Reprinted 1970, metric edition 1972, 1973
Second edition 1977
Reprinted 1978

Printed in Great Britain for Hodder and Stoughton Educational,
a division of Hodder and Stoughton Ltd,
Mill Road, Dunton Green, Sevenoaks, Kent,
by Fletcher & Son Ltd, Norwich

CONTENTS

ACKNOWLEDGEMENTS

The authors and publishers thank the copyright owners for permission to reproduce the following material:

'The Coming of the Monsoon', reproduced from *The Times* by permission.

A. T. Grove *Great Britain: Geographical Essays*, Cambridge University Press.

Gerald Durrell *The Bafut Beagles*, Rupert Hart-Davis.

Flora Thompson *Lark Rise to Candleford*, Oxford University Press.

Allen Drury *Advise and Consent*, William Collins.

Madelaine Duke *The Lethal Innocents*, Michael Joseph.

James Morris *Coast to Coast*, Faber and Faber.

Kenneth Sealy *The Geography of Air Transport*, Hutchinson.

Ed McBain *Ten Plus One*, Hamish Hamilton.

H. E. Bates *A Crown of Wild Myrtle*, Michael Joseph.

John Steinbeck *East of Eden*, Heinemann.

Mary Stewart *Thunder on the Right*, Hodder and Stoughton.

A. E. Brehrn *From North Pole to Equator*, Blackie.

C. Eyre *Vegetation and Soils*, Edward Arnold.

The authors also thank Dr. Courtenay and his publishers, G. Bell and Sons Limited, for allowing them to adapt material from the book *Plantation Agriculture*.

PREFACE

This is a book for use in the latter part of a secondary school course, preferably the fifth year, perhaps the fourth. We certainly would not advocate its use any earlier. It tries to develop overall ideas such as should arise from earlier studies. School syllabuses differ greatly, though we may assume that by the fifth year there has been modest cover of most of the world, some of it in detail. But situations vary. Pupils change forms, and staff change schools. Almost inevitably individuals have missed parts of the work. This book therefore compromises. Some new facts are offered where considered necessary to the argument, or otherwise essential. If already known this will be, of course, revision. We hope the book is self-contained, but in the main it presumes fair knowledge of three or four years work, and attempts to educe general principles. There is a constant effort to produce a world view, but detail is often drawn from the homeland.

The language is simple, but the thoughts and ideas not necessarily so. An understanding of general principles—stated or implied—should be a sound basis for examination candidates. A major weakness, in our examination experience, is incorrect logic. All too frequently candidates lack understanding of what can be legitimately deduced or argued, and of what generalisations may be made. Typical of such errors are statements that 'the Netherlands have no highlands, therefore there are no sheep' and 'as there is little lowland in Norway, most of it is bare rock'. We hope that our approach, often the analysis of the factors concerned in a problem, will help to remedy this. We have also, from time to time, indicated the correct usage of words commonly mishandled by candidates.

We claim no particular virtue for the order of the chapters. We believe that the concept of world regions is of value in school as a means of reviewing knowledge, or as a framework for its recall, and have therefore put the chapter on regions last. Teachers might well prefer to open the fifth year's work with this, to give a review of the world, and to provide background for topics studied subsequently. The rest of the book can be used in almost any order. The chapters, and many of the sections, can be handled quite separately. This should give the teacher complete freedom in using the content to support his individual syllabus.

Each section is followed by a series of exercises of varying standard. We regard them as important, and an integral part of the whole book. They further help to tie the work together, and some introduce new material which can be omitted at discretion. Some plan the basic steps in the production of simple, clear sketch-maps. Others are couched in a style popular in the newer examinations.

The unity of geography is fundamental to it, but also a teaching difficulty. Of course it must be taught in small portions. We have therefore made many cross-references, and suggest that pupils be encouraged to use them. There is also the appearance of some subject matter under more than one heading. This is done deliberately, partly as revision and more to emphasise this relational aspect of geography. But books cannot do everything, and we certainly do not suggest this one does. It merely offers an arrangement of geographical material which is becoming increasingly favoured, and which may be of assistance for the real task, which can only be done in the classroom.

M. L.
B. S. R.

Desert Duststorm ▶
(*J. Allan Cash*)

Chapter 1

PROBLEMS OF NATURE

1. Weather

'Storm forecast: reconnoitring plane centres hurricane at 26° N. 79° W. recurving north-west. Speed of advance 30 km per hour. Winds to 195 km per hour and gusts to 220 km per hour. Present movement expected to continue for next 12 hours with hurricane-force winds reaching area Palm Beach to northward 130 km during next 4 to 6 hours. Hurricane warnings hoisted from Miami to Palm Beach. Residents of low lying areas in path of centre should depart at once for higher areas. Dangerously high tides probably from Miami to Jacksonville. This is an emergency.'

You might hear this alarming type of weather forecast on your radio if you were in south-east United States during August and September, the hurricane season. Hurricanes are violent tropical storms which generally move along the tracks shown on the map (fig. 1.1). They include areas of dense cloud and very strong winds of well over 150 km per hour which blow in an anticlockwise direction towards the centre, called the eye of the hurricane. They are accompanied by torrential rainfall, which often causes almost as much damage as the winds. The winds cause huge waves which wash over houses on the coast, often drowning the inhabitants, while the wind also tears down buildings, picks up cars and hurls them down elsewhere, and causes millions of pounds of damage.

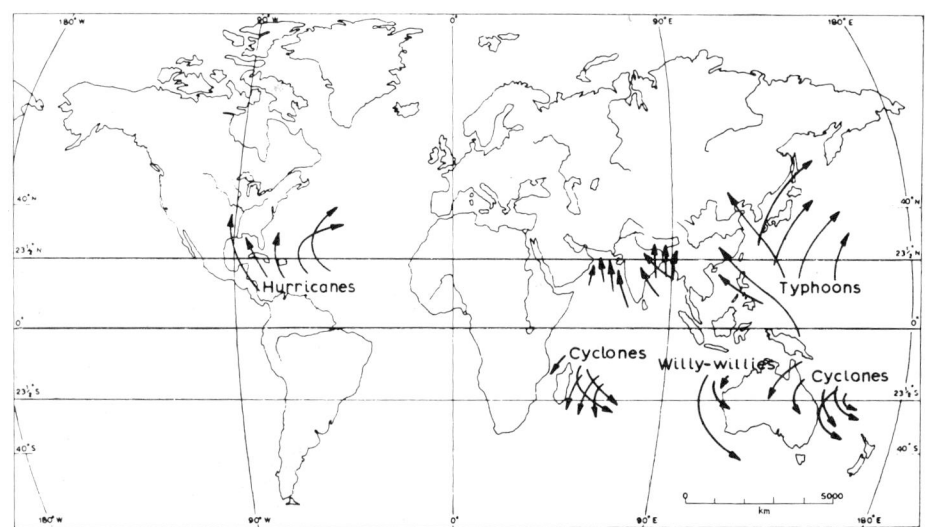

Fig. 1.1. Tracks of tropical storms.

You will see that the warning advises people to leave the hurricane lowland areas; if it is known when hurricanes are advancing precautions may be taken to minimise the damage. The storm forecast tells you that pilots fly into the eye of the hurricane to record its speed and direction. Here is an extract from an account of one such flight. 'The clouds became blacker and thicker as they flew.... As they approached the storm centre, the plane began to buck and shudder. It tossed from side to side, its wings vibrating.... The huge four-engined plane flipped here and there like a butterfly. The engines spluttered, missed and zoomed up again. The plane dropped and rose with sickening, quick changes. "In the eye of the storm, Will. Nice in here, huh?" '

The newspapers of areas where hurricanes are likely, publish advice for the public. The *San Juan Star* (Puerto Rico) of 26 August 1966 prints a whole page 'Hurricane Precaution Guide' which includes instructions such as:

1) Turn on your radio and pay strict attention to bulletins and information.

2) Store sufficient drinking water.

3) Store at least 3 days' supply of foods that do not need cooking.

4) Put in a store of paraffin lamps or candles.

5) Reinforce windows, doors and roofs.

6) Keep a first-aid kit handy.

7) Be prepared to move to the shelter assigned to you.

8) During a hurricane do not expose yourself in unsheltered places, stay in shelter, take every precaution against fire and *keep calm.*

The map shows that these storms are given different names in other tropical areas. The typhoons of Asia often cause great loss of life, ruin crops and plantations and cause extensive damage to

Miniature tropical storm: waterspout. ▶
Small cyclones over the sea are sometimes strong enough to suck up water.
(*Camera Press*)

property. In north-west Australia the storms are called willy willies and bring torrential rain in summer and early autumn. Fortunately few people live in this area, but similar storms called cyclones sometimes do great damage in the populated coasts of Queensland.

You have doubtless heard our own radio programmes interrupted while the announcer gives warnings of gales to ships. The map (fig. 1.2) will enable you to discover which sea areas the announcer refers to. Gale warnings are given when winds are expected to average speeds of 65 km an hour, or gusts of 80 km an hour. These wind speeds are much less than those of hurricanes, but such winds can blow ships off course, and cause very rough seas with big waves which can damage ships. On land gales can blow trees down and wrench roofs from houses. They are more common at sea, because the sea gives a wide open space where nothing breaks the force of the wind (fig. 1.3).

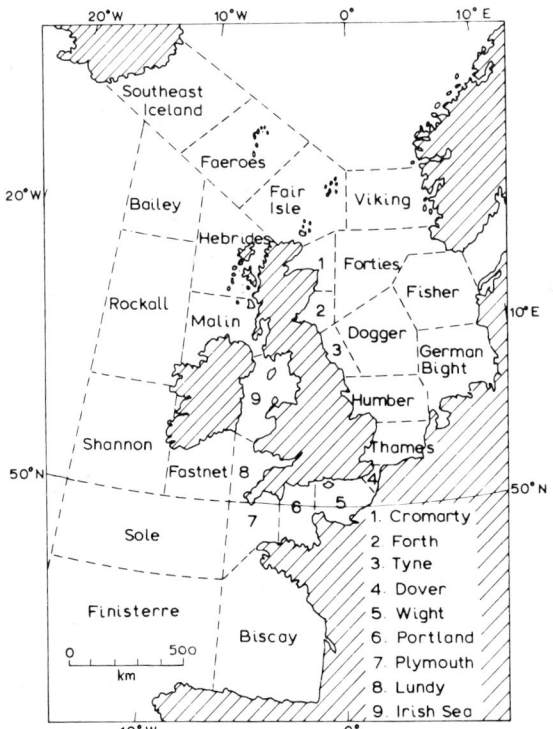

Fig. 1.2. Sea areas round the British Isles.

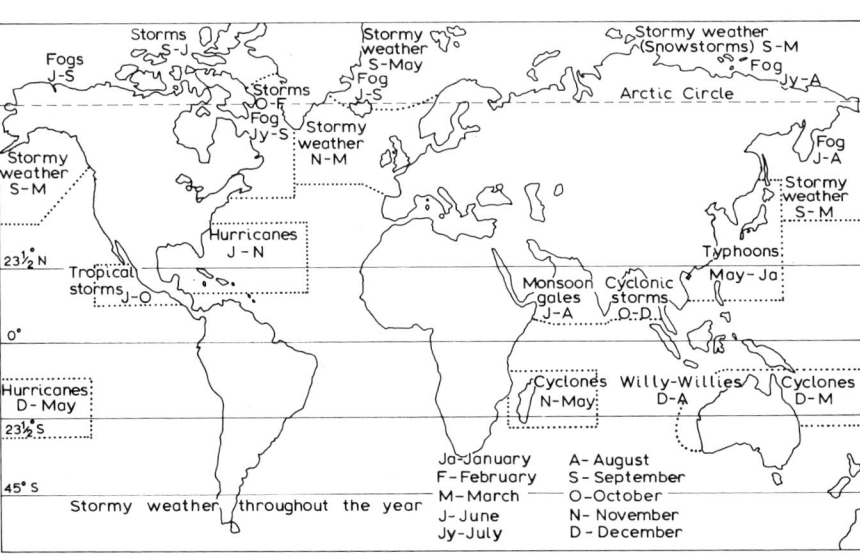

Fig. 1.3. Weather problems at sea.

Sailors and aeroplane pilots are particularly concerned with the weather, and one particular aspect of it which they dread is fog. Fog is vapour condensed in the atmosphere at or near the earth's surface. Winter fogs occur in our country when mild, damp air comes in from the Atlantic and passes over very cold ground. In rural areas fog is relatively clean, but in town the particles around which the water vapour condenses are large and dirty, e.g. soot and dust, and give rise to dirty, grey fog which may make breathing more difficult. Nowadays the exhaust fumes from continuous heavy traffic add to the impurities in the air. The word 'smog', made up from the words smoke and fog, is used as a name descriptive of the dirty yellowish-grey vapour which may engulf vast urban complexes. Los Angeles, with its great aeroplane, car, oil and other industries and extremely heavy traffic suffers a great deal from smog. London has fewer smogs these days. Smokeless zones, in which no one may burn fuel which gives off smoke, coupled with the compulsory tall chimneys of factories, from which smoke is likely to be dispersed by wind before it can become harmful, have lessened the number of severe smogs.

If smog or fog occurs over landing grounds pilots have to take their planes to another fog-free airport. This may be very inconvenient for both crew and passengers. Ships may run into danger in sea fogs which occur where warm sea currents meet cold sea currents. For example, the warm air over the Gulf Stream is chilled when it passes over the Labrador current, and the resulting fogs hinder fishing round the Grand Banks of Newfoundland. Fogs often occur on coasts where warm air from the sea meets colder air from the land. Nowadays most large ships are fitted with radar, which tells them when other ships or lands are near by, but such equipment may be too expensive for small vessels. Fog is indeed a great hindrance to all forms of transport. It forms only when the air is still, for wind disperses it, or, more fortunately, prevents its formation.

Among the people most directly affected by weather is the

Fog over rural England.
(*Aerofilms*)

farmer. He plants crops which are likely to flourish in the climate of the land where he lives, but he worries about the possibility that the weather may be different from normal. Insufficient rain in the growing season may prevent plants from growing strong and healthy, or cause seedlings to die. Heavy rain when wheat, oats or barley are ready for harvesting may flatten the cereals so that it is difficult or impossible to cut them. A late frost, that is a frost which occurs when temperatures are normally at that time of year much higher, may kill the blossoms on fruit trees so that no fruit forms.

There are many plants, and even trees, which die in heavy frost,

5

so that it is important that you should know why frost occurs. To understand this properly, you must know something about the atmosphere. Air normally contains moisture. This is water which has been changed by evaporation from its liquid state into vapour. You see this happening when a kettle boils and clouds of water vapour are discharged into the air. Warm air can hold more water vapour than a similar volume of cold air. Electric hair dryers are based on this fact, and produce a blast of warm air which absorbs the water from the wet hair. If air is cooled it loses some of its capacity to hold water; the air becomes saturated and some of its vapour condenses into liquid again. You see this when the kettle boils in a cold kitchen; the walls become wet and may even stream with water which has condensed from the water vapour pouring from the kettle spout. In nature, dew is formed by condensation on cold solid objects like buildings, plants or the surface of the ground. Frost occurs when the condensed water is frozen; this happens when temperatures are at or below freezing. Rapid freezing produces the white frost which makes trees and plants look so attractive; slow freezing produces clear, glassy ice.

In general, plants which are likely to be harmed by frost are sown only when warm temperatures are well established. Citrus fruits, the lemon, orange and grapefruit, are very sensitive to frost, and are grown mainly in lands with a Mediterranean climate (page 187) where winters are mild. Even so, unexpected frosts in the Central Valley of California and in parts of Florida have caused great damage to the citrus groves. Nowadays farmers keep rows

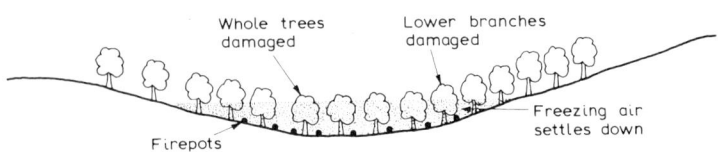

Fig. 1.4. Frost prevention.

of fire-pots in their orchards. These are metal containers filled with fuel ready for lighting as soon as temperatures fall below freezing (fig. 1.4). When lit they warm the air so that frost does not form, and furthermore, so that the moisture in the plants themselves does not freeze. Since weather forecasts have been available, the farmer knows when cold spells are coming, and his thermometers are fitted with an alarm which sounds as soon as the danger point is reached. He always hopes that he will not need to light the fires, for this involves much labour and is very costly.

Nowhere in the world are temperatures too high for plants to grow, providing that they have sufficient water. Where rainfall is known to be scanty, it is often possible to provide water by irrigation (page 48). There are, however, some areas of the world where the rainfall in most years is sufficient for crops, but in some years it is not. It is in these areas that famine is likely to occur when crops fail, that is when they die or do not grow properly. Among such areas are parts of India relying on monsoon rainfall which, due to arrive in June, may not come, or comes much later and does not last long enough. 'For 15 days or perhaps more one waits in expectation; after 20 days the expectation deepens into anxiety, and as the days go on and the sky is still clear, with a horrible relentless blue, anxiety becomes despair. There is no more depressing sight than that of the young rice, wilted and dying, with the pitiful husks that refuse to fill in the baked fields where water should be standing centimetres deep.' Scientists in various countries have attempted to produce rain. They know how to make clouds discharge their moisture so that rain falls, but the process is costly and not always successful, even over a small area.

Since we do not know what tomorrow's weather will be, we have to rely on weather forecasts. These are so familiar to us that we tend to take them for granted, and seldom think of the vast organisation of people whose work ensures their accuracy. All over the world there are weather stations, where trained observers record barometric pressure, air temperatures, the humidity of the air, the direction and speed of wind, the height, amount and type of cloud, rainfall and visibility several times a day. These and other findings, often sent by radio, are put together in various government meteorological offices, and recorded on maps. By their study experts are able to forecast weather. In countries like ours, where the weather is always changing, it is possible only to give details of what the weather is likely to be for a day or so ahead. In countries where the weather remains the same over long periods forecasting is easier. The advent of the aeroplane has greatly assisted forecasting, for pilots can give details of conditions in the upper atmosphere. Knowledge of the great air masses over our lands and those of the Polar and Equatorial regions is vital, since they affect our weather.

We have seen that weather information is vital to farmers, flyers and sailors. Warnings of icy roads and persistent fog are also broadcast for the benefit of all travellers. Even in this highly scientific age, little can be done to control weather, but foreknowledge of it means that some steps at least may be taken to minimise its bad effects. There is need for information about weather all over the world, and all countries co-operate to provide it. This international co-operation is a splendid example of how countries can pool information for the benefit of the whole world.

Work to do

1. *'Tough new smog law is likely to improve the health of Californians at the expense of their pockets. Every new car sent to the Golden State must be checked to ensure that its smog-control device works. Detroit car-makers will take care of the inspections but they say it means car prices going up by at least £10'* (Daily Express, *31 July 1968*).

i) *What is meant by smog?*
ii) *Why is it harmful to health?*
iii) *Why do you think Californian cars must have a smog-control device?*
iv) *Where do California's cars come from? Find the town in your atlas and name the state in which it lies.*
v) *Why do you think California is called the Golden State?*

2. *Study the maps* (figs. 1.1 *and* 1.3).

i) *Make a list of countries where typhoons occur.*
ii) *Write a brief description of a typhoon.*
iii) *What are storms called in* (a) *Australia and* (b) *Malagasy* (*Madagascar*)?
iv) *Which parts of the world suffer from severe sea fogs?*
v) *In which latitudes is there stormy weather throughout the year? Why are these storms less harmful to man than they might be?*

3. i) *What is frost?*
ii) *Why does it form?*
iii) *Why is it sometimes harmful?*

4. *Write a detailed description of today's weather, under the headings: sky, cloud cover, wind direction and strength, amount of sunshine and/or rain, temperature, humidity (dampness or dryness of air). Add any other details which you find of interest.*

5. i) *What is meant by a weather forecast?*
ii) *Which people find weather forecasts particularly useful?*
iii) *Which aspects of weather give rise to problems for man? Write a sentence on each to describe the difficulties found.*

2. Soil conservation

Soil is one of the world's most important natural resources. Soil has been defined as 'the upper layer of the earth's mantle of rotted and broken rock, which has been changed by the action of living things such as plants, animals and bacteria' (microscopic organisms found almost everywhere). It is perhaps easier to think of soil as a mixture of particles of weathered rock and decayed vegetable matter or humus in which bacteria are active (fig. 1.5). The nature of the soil in any area is dependent on the underlying or parent rock, and on the climate and vegetation. In hot deserts like the Sahara the soils are sandy because they consist almost entirely of particles of parent rocks, for there is little vegetation to provide humus. In cold deserts like the tundra they consist mainly of shattered rocks, so that they are very stony and similarly lacking in humus. Some of the most fertile soils, like the black earths of the Russian steppe, consist almost entirely of humus in their topmost parts. Humus is rich in plant food. Sometimes man adds humus to the soil by ploughing in a crop. Cotton plants are sometimes ploughed in when there has been overproduction; a crop of peas or beans which are useless for picking after heavy rains or a season too dry for their full growth is not entirely wasted if ploughed back into the soil, while grain stubble is normally turned back into the land during the next season's ploughing. Nature's way of adding humus to soil is through falling leaves and dead plants which rot down as they lie on the ground. Thus a cover of vegetation is very important in maintaining soil fertility. It is equally important, as we shall see, in preventing soil erosion.

Erosion is a natural process. All land surfaces are attacked by the various forms of atmospheric weathering. Mountain peaks and ridges are shattered by frost; bare rock is split by the heating of the sun's rays or by sudden cooling at night; water freezing in cracks of rock expands when transformed into ice, and so enlarges the cracks; heavy rain cuts gullies in hillsides and moves particles

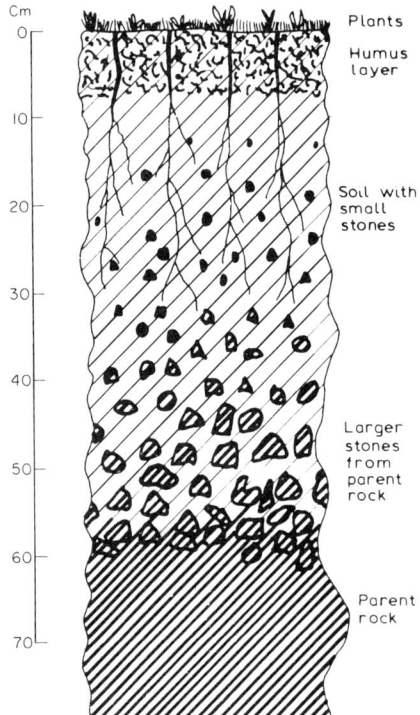

Fig. 1.5. Soil diagram.

of earth, while winds blow loose particles of soil away. If soil is unduly exposed to the elements of wind and rain these natural agents of erosion will soon remove it, eventually leaving the parent rock exposed.

A good plant cover helps to prevent loss of soil. The roots of living plants help to hold the soil particles together. Many soils are porous, and rain soaks or percolates into them, but heavy rain may close up openings in the surface and so run off instead of into the soil, carrying with it soil particles. A leafy covering checks the force of heavy rain, and run off is less severe. Plant cover also hinders rapid evaporation, which would dry and loosen the soil. It also breaks the force of the wind at ground level, thus preventing it from carrying off or transporting soil particles.

Man has not always borne in mind the protective nature of plant cover. In spreading cultivation and stock rearing he has ploughed up natural grasslands and cut down huge areas of forest. Ploughing exposes bare earth to the elements, so does leaving land fallow (page 70). If too many animals graze on grass it is cropped so short that it ceases to be a natural cover. Sheep crop particularly closely; so do goats. Furthermore, they eat the seedlings of trees and shrubs, so destroying the natural replanting process. As a result, many hill slopes have been entirely stripped of trees, and on these slopes wind and rain do their worst. In short, man has been, and to some extent still is, thoughtless in exposing large areas of land to soil erosion.

In 1935 the area of the middle west of the United States, particularly Kansas, Oklahoma and north-east Texas, experienced severe drought. The drought continued for several years, during which the top-soil became so dry that it was easily blown away by the wind. In Oklahoma one large area is estimated to have lost some 50 million tons of soil in a single day. In Texas there are still houses which lie completely buried beneath rich black soil windblown from farms to the west. The area, once fertile, became known as the 'Dust Bowl' (fig. 1.6). The development of the 'Dust Bowl' occurred through sheet erosion, that is the clearing of large areas of soil largely by the wind. Rainfall is also an agent of erosion. Violent rainfall, such as often occurs in thunderstorms, will run rapidly down any slope or depression lacking vegetation, cutting itself a channel. This channel attracts water down both its slopes; soon branching channels result. These channels are called gullies, and this type of erosion is gully erosion. It is particularly disastrous in sloping or hilly land, and has been the cause of much loss of soil. In some areas of bad gullying the name 'bad lands' is

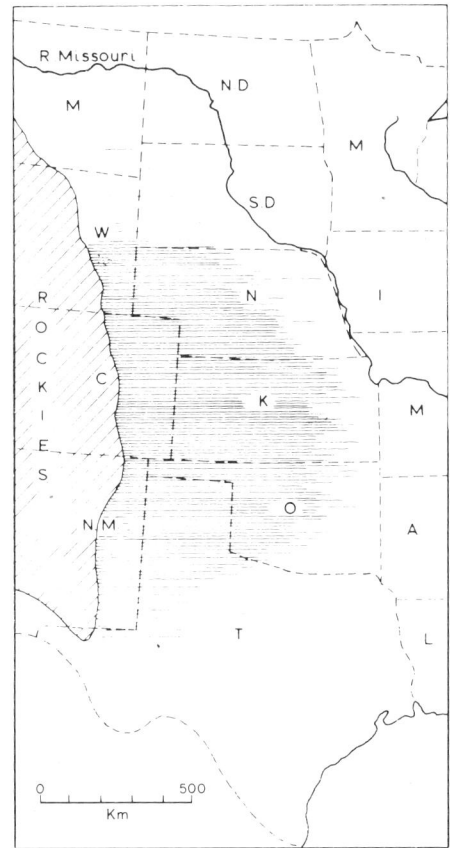

Fig. 1.6. Areas affected by soil erosion: the 'Dust Bowl'.

used. The soil removed by rain may get washed out to sea in rivers. In floods it may spread a cover of debris over the flood plains. It may get carried into reservoirs and start filling them up, reducing their water-holding capacity.

9

◄ **Gullying: Tennessee, U.S.A.**
Draw your own annotated version of this picture.
(*U.S. Information Service*)

Sheet and gully erosion threaten crop growing in so many countries that here we will select only a few examples. It is thought that the Missouri Valley of the United States has lost 20 cm of top-soil in the past thirty years, while the Mississippi River deposits over 400 million tonnes of mud yearly into the Gulf of Mexico. In South Africa 30 per cent of soil fertility has been lost, gullying being especially marked in the Karroo lands. In Kenya and Rhodesia overgrazing has caused soil loss in large areas. In Mediterranean Europe sheep and goats have cropped mountain slopes so closely that much land has been ruined. In New Zealand during the past century three-fifths of the native forest has been cleared for cultivation, and heavy rain has removed soil from the unprotected slopes. Almost one-quarter of the entire country is affected by soil erosion. During the Second World War many of the forests of Indonesia were cut down; Java particularly suffered from the resulting gully erosion. Indeed, soil erosion is now recognised as a world problem.

Now that man recognises the problems of soil erosion, in many countries he is trying to prevent it or stop it spreading further. The Government of the United States has a Soil Conservation Department, which gives advice not only to American farmers but to those of other lands. This department has experimented to find all possible ways of conserving the soil, and selects those appropriate for each particular area. The picture shows contour ploughing; this method of ploughing along the contours of the land checks the downward flow of run-off. Strip cultivation is encouraged; here grass and crops, or different types of crops, are

Contour ploughing: Pennsylvania.
◄ Draw this picture as a field sketch and then add contour lines
(*U.S. Information Service*)

Shelter Belt: Kansas.
How does the scenery in this picture illustrate the information on page 11?
(*U.S. Information Service*)

planted in alternate strips. The crops ripen at different times, so only single strips, rather than large areas, lie empty after harvest, and these are replanted quickly. Soil erosion on slopes is checked by the planting of trees, called re-afforestation, or by making terraces edged with retaining walls which check the downrush of rain-water. Windbreaks of trees are planted as shelter-belts against wind. Long-rooted grasses are also planted to bind the soil together in areas subject to wind erosion. In gullies little dams of brushwood are laid across the V-shaped depressions; grass is then allowed to grow, or shrubs or vines are planted. These not only bind the soil and help soil to collect but provide humus. Eventually enough soil may be restored to make it worthwhile to add fertiliser and plant crops.

As well as the efforts made by individual farmers, many governments have regional plans for preventing soil erosion and for reclaiming land attacked by it. Probably the first of these was that of the Tennessee Valley Authority in the United States. The lower slopes of the Appalachian Mountains, through which the Tennessee River flows, had long been stripped of forest. Not only were the farms in the valley very poor but the river itself often flooded violently, causing great damage. The T.V.A. controlled the river by building many dams, produced hydro-electric power which enabled the development of industries and raised the standard of farming. They persuaded the farmers that no more trees should be cut down, that forests should be replanted, that they should use contour ploughing instead of ploughing straight down slopes and that land should be fertilised together with the use of crop rotation. The hydro-electric power also produced cheap fertiliser from the nitrogen in the air. The Tennessee Valley is now a prosperous farming and industrial area. In Australia the Murray Basin has been seriously affected by soil erosion; the states of New South Wales and Victoria, with the co-operation of the farmers, have developed plans to ensure irrigation water for dry areas, for moist soil is less subject to wind erosion. The plans are for the control of the waters of the Murray River and its tributaries. In the arid steppe land of European Russia there is a fifteen-year plan to replace the grain–fallow system (page 70) by a grass–grain rotation. The map (fig. 1.7) shows you the vast shelter belts which are being planted as protection against the strong winds. Extensive irrigation systems are also being constructed.

It is not easy to persuade man to adopt soil-conservation methods. Sometimes no one is aware of sheet erosion until the top-soil has already been lost. Some farmers are aware of loss of soil through erosion, but need to be taught how to prevent it. It is difficult to explain to African or Indian peasants how grazing too many cattle causes soil erosion, or that soil fertility may be restored if different crops are planted in rotation instead of the same crop every year. In any case, soil conservation often

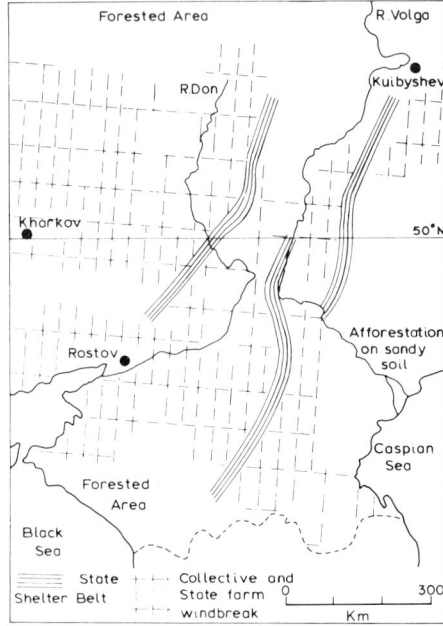

Fig. 1.7. Shelter belts: U.S.S.R.

requires much labour and can be expensive, especially if carried out on a large scale. Nevertheless, it is essential that man takes active steps to guard the soil, since it is the source of his food supply. Organisations such as U.N.O. have teams of specialists to tackle soil erosion, and countries co-operate on a world-wide scale in the fight against this particular problem of nature.

Work to do

1. *Read the following passage, then answer the questions:*

   Parts of North America have been farmed carelessly, for large open treeless areas have been ploughed and cropped, and insufficient humus replaced in the soil. When for several successive years the annual rainfall was lower than average, the soil became drier and drier, eventually turning to dust. In these areas the summer winds caused great harm. They swept the dust away, sometimes for hundreds of kilometres, so that the land was left with no top-soil, infertile.

   i) *Name the problem described in the passage.*
   ii) *What natural agent caused it?*
   iii) *Describe the type of land where this problem occurred.*
   iv) *What caused the soil to become drier and drier?*
   v) *Name any states in North America where these conditions may occur.*

2. i) *Name a natural agent other than wind which removes soil.*
   ii) *On what type of land is this agent particularly destructive?*
   iii) *Name any one part of the world where you know such destruction has occurred.*
   iv) *State the difference between sheet and gully erosion.*

3. *Give reasons why a good plant cover helps to prevent loss of soil.*

4. *Write a paragraph describing how man tries to prevent soil erosion.*

5. *Copy the soil section* (fig. 1.5). *Using the diagram to help you, write a description of the soil shown.*

3. Pests and diseases

'King Rats! There are over 50 million in Britain.' Thus runs a headline in the *Evening News* of 18 April 1966. The article goes on to say how the Ministry of Agriculture deals with these pests, and adds, 'You can treat all the farms and their buildings, but it is still difficult to tackle the rats in the hedgerows and ditches.' We seldom think of our own country as having pests, but farmers would include squirrels, jays, wood-pigeons and other birds among them. There are also diseases which attack farm stock and are feared because they spread so rapidly. You may have heard B.B.C. announcements concerning outbreaks of fowl-pest, or foot-and-mouth disease among cattle. The outbreak of foot-and-mouth disease which occurred in England necessitated the slaughter of 444 000 cattle, pigs and sheep worth £27 million late in 1967. The animals were slaughtered in an attempt to stop the disease spreading. This slaughter is still necessary, even in countries like our own, where there are many veterinary surgeons and medical science is very advanced.

If we consider the above facts it is easier for us to understand the difficulties experienced over the whole world where other pests and diseases are prevalent. In all countries where potatoes are grown the little, striped Colorado beetle is feared, because it can destroy the crop. In the Cotton Belt of the United States (fig. 1.8) attempts to destroy the cotton-boll weevils cost thousands of dollars. These are small beetles, millions of which hatch out from eggs in the spring. They live on the cotton plants and eat through the cotton boll, making the threads of cotton too short to be usable, or destroying them entirely. A hard frost kills the beetles, but a mild winter and a wet spring favour their survival. The Americans try to cut down the scrub where the beetles breed in the neighbourhood of the cotton plantations, but they cannot remove it all. The cotton fields are also sprayed, either by hand or from aeroplanes, with an insecticide which kills the weevils. Flying the

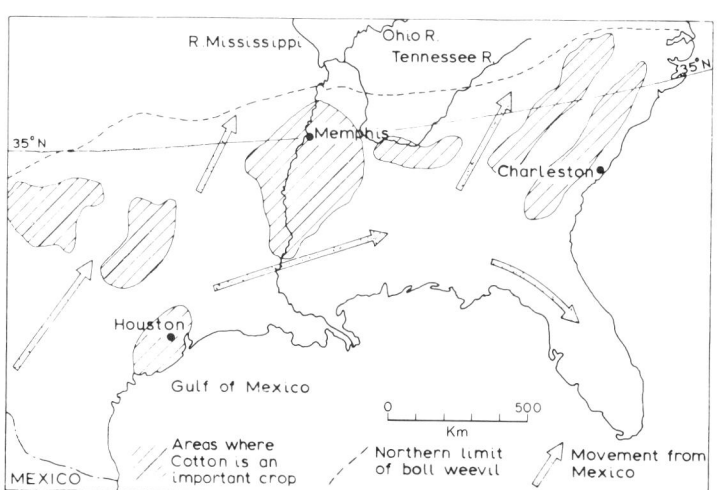

Fig. 1.8. The spread of the boll weevil.

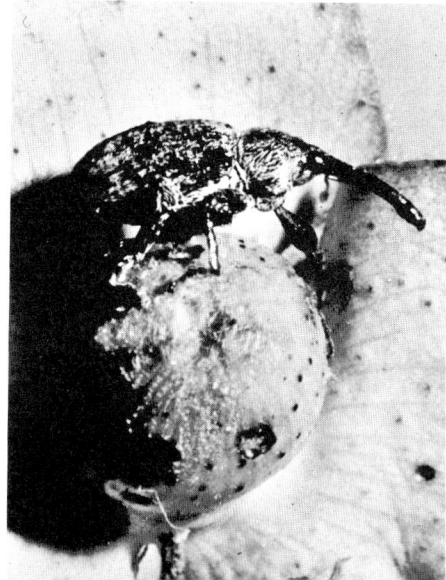

Boll weevil.
The dark spots on the cotton boll show the damage caused by the beetle.
(*Shell*)

Spraying insecticide over a sugar plantation: Venezuela.
(*Shell*)

Desert locust.
(*Shell*)

aeroplanes at very low levels to make sure that the spray covers the plants is a very skilled, dangerous and therefore highly paid occupation. Despite all efforts, much cotton is lost to the boll-weevil.

The locust, which is like a large winged grasshopper, is one of man's oldest enemies. It is well known in the tropical lands of Asia and Africa. Locusts were the seventh plague of Egypt in the time of Moses, and 'very grievous were they; for they covered the face of the whole earth, so that the land was darkened; and they did eat every herb of the land, and all the fruit of the trees; and there remained not any green thing in the trees, or in the herbs of the fields, through all the land of Egypt' (Exodus x, verses 14 and 15).

Tropical Africa suffers from three types of locusts: desert locusts, red locusts and the migratory locust. The map (fig. 1.9) shows where they spread. When young they are called 'hoppers', and travel along the ground, destroying everything green within their path. When they fly the swarms may travel at several hundred metres above the ground at a speed of about 16 km an hour. Scientists have found that there are certain breeding grounds from which swarming takes place periodically. In 1926 small swarms of the migratory locust were seen by the River Niger at Timbuctoo. By 1929 the whole of West Africa had been invaded; by 1930 the locusts had reached Egypt, and by 1932 they had reached Rhodesia. Modern insecticides can protect an area of crops reasonably well, but it is difficult to control a large-scale attack. Aeroplanes are used to fly in front of swarms on the wing, spraying them in the air. Scientists are at work finding ways to destroy the breeding grounds. The experts of the World Health Organisation now make monthly maps to show where the locusts are, and try to forecast where they will swarm. Many millions of pounds have been spent in the fight against the locust, and it is likely that more millions will be necessary.

The house-fly is a common pest in Europe. It is a dangerous insect because it can spread dirt and disease, but the use of

14

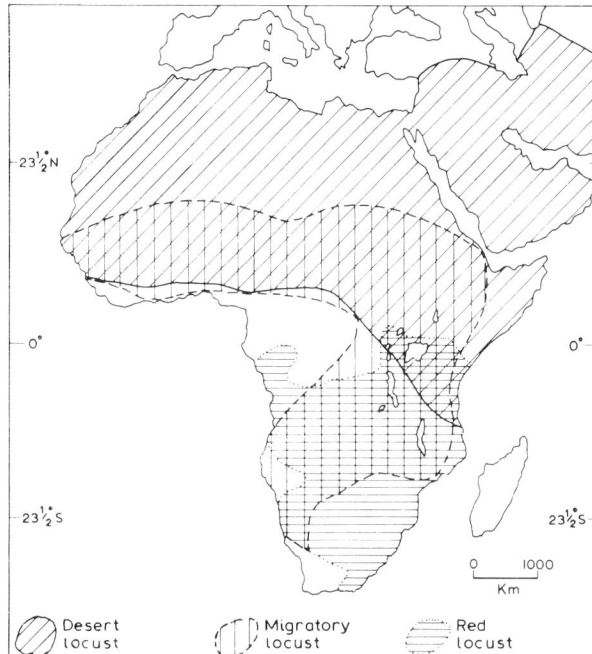

Fig. 1.9. Locust areas in Africa.

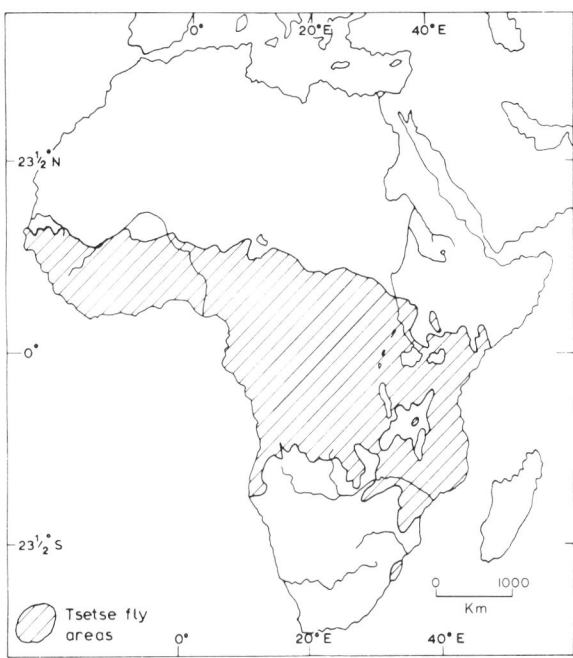

Fig. 1.10. Tsetse fly areas in Africa.

insecticides keeps its numbers down. In Africa the name tsetse fly is given to some twenty species of blood-sucking flies which are larger than house-flies. Some of these flies transmit diseases by infecting the man or animal they bite. To man they transmit sleeping sickness, which can now be cured by drugs if quickly available. To cattle, horses and dogs they give a disease from which the animals usually die. To get rid of them it is necessary to clear away vegetation, especially that surrounding villages, but the flies lay their eggs in shady places, so that they are almost impossible to locate. The map (fig. 1.10) shows the vast areas infected by these flies.

The mosquito is not unknown in our country, and we mention its presence in the Arctic lands on page 209. Fortunately, the bite of these insects, though irritating, is not usually serious. Bites from the anopheles mosquito of the tropics, however, carry germs of one of the most widespread diseases, malaria. Quinine (page 174) has long been used to lessen the severity of malarial attacks. Nowadays more powerful medicines are available, and people visiting the tropics normally take a small tablet of atebrin or mepacrine daily, which seems to produce immunity. Nevertheless, these drugs are not available all over the vast tropical lands yet. It is clearly most sensible to try to get rid of the mosquito. To do this, attempts are made to prevent it from breeding. The mosquito deposits its eggs where there is stagnant water, especially in marshy areas. A thin film of oil sprayed over the water surface prevents the larvae from breathing. It is possible to spray water surfaces round towns and villages, but enormous areas are left unsprayed. It is an offence in some tropical countries to leave any tin cans or other vessels lying about in which mosquitoes can breed. Other types of mosquito carry diseases such as yellow fever. For all travellers to the tropics inoculation against this disease is compulsory. Inoculation is not yet available to all who live in the tropics.

There are many other disease-carrying insects in the tropical lands. The three we have described will suffice to show the great problems they create. Crops may be eaten by locusts; cattle cannot be kept in tsetse areas, and man may be prevented from working by bouts of malaria. Many doctors, nurses and scientists who work for the World Health Organisation are trying to cure the diseases and destroy the pests. Nevertheless, these lands are not all healthy lands, and the inhabitants need help in trying to make them so. There is much work to be done the world over in the fight against disease.

Work to do

1. *On an outline map of Africa mark in the path of the migratory swarms of locusts described on page 14. Give your map a key and a title.*

2. i) *What insect carries malaria?*
   ii) *How does man attempt to destroy the insect?*
   iii) *What drugs can be taken to:* (a) *prevent;* (b) *alleviate malaria?*

3. i) *Make a drawing of a locust.*
   ii) *Explain why locusts are pests.*
   iii) *Using the map* (fig. 1.9) *and your atlas, make a list of the countries which are liable to locust attacks.*

4. *Write verses 14 and 15 from Exodus x (page 14) in your own words.*

5. *Using this section and section 4, make a list of all the insect pests. Write a sentence on each to describe the damage it does. Add any other insect pests known to you.*

4. Forests

The map (fig. 1.11) shows the distribution of the world's forests. The key tells you that there are are two main types of trees, softwoods and hardwoods. The softwoods are usually evergreen trees with needle-like leaves, such as larch, spruce, pine, balsam and cedar; these are conifers. Hardwoods have broad leaves. In temperate lands they include oak, chestnut, maple, poplar, hickory, ash and elm; in tropical lands the many varieties include ebony, mahogany, rosewood and purple heart (page 172). As you see from the map, the great forest belts of the world are the coniferous forests which extend across Canada, Europe and Asia roughly between latitudes 50° N. and 70° N. (page 197) and those in the equatorial regions of South America, Africa and Indonesia. Between these lie many other forest areas. For example, one-third of the United States is forest land. Nearly every state has some forests, and nearly 1000 different kinds of trees grow therein. Of these only about 100 varieties are used in lumbering.

The main problem of forests lies in the need to conserve them so that the timber supplies of the world do not become exhausted. The forests of the world were originally much more extensive than they are today, particularly in the temperate latitudes. For example, the natural vegetation of the British Isles is basically forest. If left to nature, the only large areas which would not be forested are those where there is too little soil for trees, those which are too dry because of highly permeable soil and those where poor soil drainage leaves the land too wet. The once vast forests of the Scottish Highlands, the Southern Uplands and Wales have long since disappeared, save for small scattered remnants on some valley sides. The forests were largely removed by charcoal burners and sheep graziers; they were also felled for shipbuilding, fuel, buildings and to clear land for agriculture. Now only some 5 per cent of our total land area is woodland, and much of this has been planted by man. In Europe and the United States great destruc-

Fire precaution notice: Victoria, Australia.
(*Australian Information Bureau*)

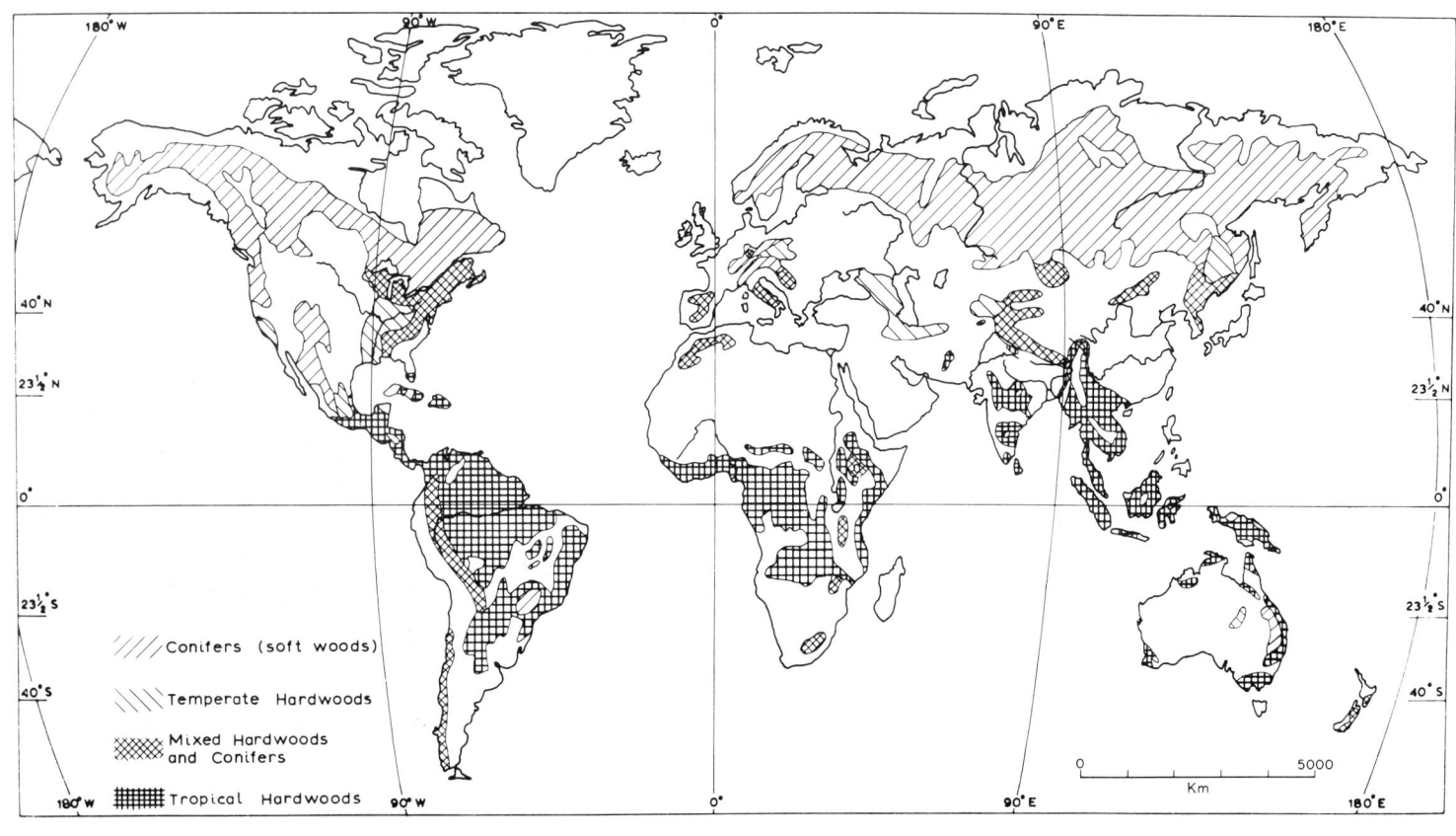

Conifers (soft woods)

Temperate Hardwoods

Mixed Hardwoods and Conifers

Tropical Hardwoods

Fig. 1.11. Distribution of the world's forests.

tion of forest has also taken place. In the United States early settlers were encouraged to clear away trees not only for house building and fuel but to enable land to be cultivated.

Although the great equatorial forests of the Amazon stand virtually as they have always done, the practice of shifting cultivation has altered the type of vegetation over wide areas of South-East Asia. In order to clear the land for growing crops the original forest is fired. When the cleared land is abandoned after two or three years it does not always return to the original rain forest but often grows back as open grassland. In the Philippine Islands, for example, about one-fifth of the total land area is now grass covered where once grew forests. In Borneo too large quantities of virgin forest are destroyed every year by shifting cultivators. In Java and Sumatra large forests have been lost to permanent agriculture.

Man is not the only agent of destruction in forest land. Many young tree saplings or sprouts are eaten by grazing animals, both wild and domestic. Trees are subject to disease. In Pennsylvania an insect blight necessitated the cutting down of a huge area of chestnut trees as being the only way of controlling this slowly spreading disease. Even so, the trouble lingered for forty years. Cocoa trees have been destroyed by 'swollen shoot'; other trees can suffer from 'Dutch elm' disease. This disease is carried by the European bark beetle. Clearly insect pests also attack trees. In Oregon, for example, the 'hemlock looper', a moth caterpillar, started to cause serious damage in a large area of forest. It took some time for scientists to find a powder which would kill the caterpillar but be harmless to other wild life, and nearly 1000 low-level aeroplane flights were made to dust the powder over the affected forest. In Ontario many thousands of square km of valuable timber has been destroyed by the bud worm, the larva of a moth which lays its eggs on the buds of spruce and balsam trees. When the caterpillars hatch they eat the tree buds, thus destroying growth. In fact, in forests nature herself seems the next deadliest foe to fire. The problems of forest fires are discussed on page 205.

Fire precaution watch tower in the Rockies.
(*U.S. Information Service*)

Cause	*Percentage of all forest fires in the United States*
Careless smokers	25·5
Deliberate firing	24·8
Débris burning	13·6
Lightning	9·1
Campers	6·6
Railroads	4·2
Lumbering	1·9
Other causes	14·3

The table serves to remind you of some of the causes of such accidental fires. You can draw a bar-graph to illustrate these figures. The loss of timber through human carelessness is very great. British Columbia spent over $4 million fighting 3102 forest fires in 1961. In Ontario, which averages 1400 forest fires each year, the Government maintains a summer fire-fighting force of 2000 men, 300 look-out towers and 27 air bases for parachute fire-fighters and water-carrying planes. Even so, it is estimated that 8 square km of forest are destroyed by fire every day. No wonder that in our own small forests notices beg us to be careful not to cause fires.

In the equatorial forest lumbering is limited by the fact that different species of trees grow individually and not in stands (page 173). Nature largely replaces the trees which are cut down by man. In other forests lumbering has largely become much more scientific in order to conserve timber as much as possible. Trees are selected for cutting according to their age, size and the purpose for which they are required. Much less wood is wasted than previously (fig. 1.12). Tree-trunks may be used for telegraph poles and ships' masts. The saw log or main trunk may be converted either into construction lumber for buildings or into industrial lumber for furniture, machine equipment or containers of many kinds. Some timber is now converted into plywood, which is very strong: it is used, among other things, for furniture, in aeroplane construction, for prefabricated houses and packaging. Nowadays even shavings, sawdust and other oddments of wood are used to make wood pulp. The use of pulpwood in the manufacture of paper is discussed on page 204. Wood contains about two-thirds cellulose and one-third lignin, and pulpwood is largely cellulose. Cellulose is also used in making plastics, alcohol, gunpowder and photographic film; lignin is used for tanning leather, as a water softener and as a binder for mixing concrete. The bark of trees is now used to make special lubricants and plastics. Even the waste liquid and gases from pulp mills are converted into chemicals. Wood is indeed a valuable natural resource. You can

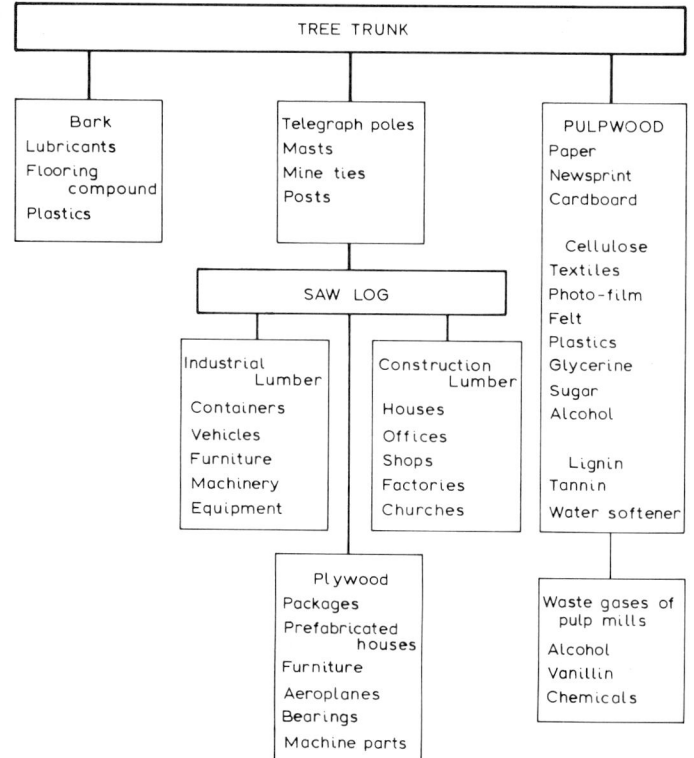

Fig. 1.12. Products from timber.

make a list of all the things you know to be made of wood, naming, where possible, the wood used.

Now that the value of forests is appreciated by man, many countries are trying to conserve them. The idea of forest conservation involves two main considerations; first, that more timber should be grown than is cut down; secondly, that forest should be allowed, where possible, to renew itself by natural seeding. Where natural seeding does not take place readily, trees are planted. This tree planting is known as reafforestation. It may take between twenty and eighty years for trees to grow to full size, so man has to be patient. He is now replanting trees for future generations to use.

The aim of our own Forestry Commission, established in 1919, is to improve some 800 000 hectares of existing woodland and to plan an additional 1 200 000 hectares, in order to supply 30 per cent of our normal need for timber. At present we import 90 per cent of our timber. Already 400 000 hectares have been planted in

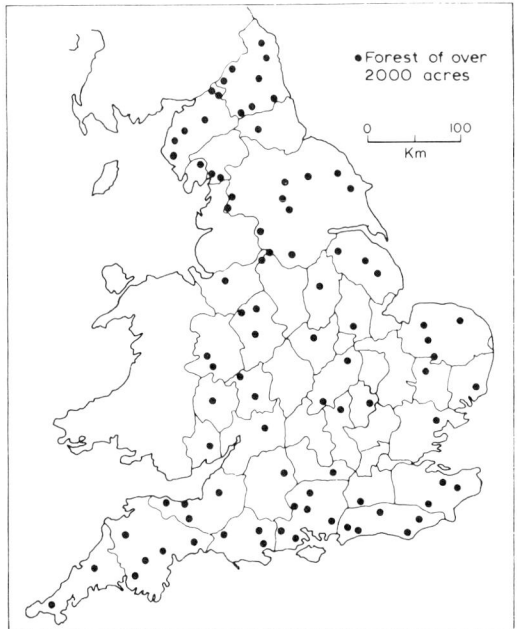

Fig. 1.13. Forestry Commission: main forests.

Reafforestation: Highlands of Scotland.
Note the great change in scenery from open moorland to forest. A problem is that walkers complain open country is lost to them.
(*L. J. Long*)

Great Britain, together with 16 000 hectares in Northern Ireland and 80 000 hectares in Eire (fig. 1.13). Since most of the wood needed for industry is softwood, most of the trees planted are conifers, mainly spruce and larch. You may have seen areas of young 'Christmas trees' planted closely in rows on land which was previously used for rough grazing. They are commonly found on the Scottish Highlands, the Pennines and in Wales. The oceanic climate of our islands ensures rapid growing, so that timber large enough for pit props can be thinned out after fifteen years, while felling for saw timber may take place in as little as twenty years.

Not only are new forests being planted but existing woodlands are being improved. Unhealthy or crowded trees are cut down, undergrowth is cleared and new trees are planted.

You will see that the main problem of forestry is the conservation of trees. Trees are valuable not only for timber but sometimes because their sap is useful; for example, pine gives turpentine and resin, the maple maple syrup, and rubber latex. Trees can be destroyed by fire, disease and insects and by grazing animals. Man is trying to overcome the loss of timber by tree care and reafforestation. Most of the trees replanted are softwoods, because they mature quickly, but some hardwoods are being planted. Man is also trying to make full use of all cut timber. The timber industries of tropical lands are growing, and many unfamiliar varieties of tropical hardwoods are now used in our country, imported from the Commonwealth. Forest conservation, practised now in many countries of the world, provides the hope that man will be able to maintain timber supplies in the future.

Work to do

1. *The following table shows the main countries of the world producing wood pulp:*

Country	Million tonnes yearly	Country	Million tonnes yearly
U.S.A.	38	Finland	6
Canada	17	U.S.S.R.	6
Sweden	8	Norway	2
Japan	8	Germany (West)	1

  i) *Draw a column graph to show this production, using graph paper and a vertical scale of 1 small square to 1 million tonnes.*
  ii) *What type of timber is used to make wood pulp?*
  iii) *To what uses is wood pulp put?*

2. *For what reasons have the forests of the world been so largely destroyed? Give real examples to illustrate your answer.*

3. *Study the map of Forestry Commission forests in England* (fig. 1.13), *and answer the questions:*

  i) *Which county has the largest forest areas?*
  ii) *Name the counties which have no large conservancy schemes. Can you suggest any reasons for this?*
  iii) *Name a forest in* (a) *Hampshire* (b) *Nottingham.*
  iv) *Why are there so many conservancy schemes in the Lake District?*

4. *The following table shows new factories in the United Kingdom using wood as raw material:*

Location	Product	Location	Product
Annan	Chipboard	London	Fibre board
Inverness	,, ,,	Queensferry	,, ,,
Hexham	,, ,,	Fort William	Paper pulp
Thetford	,, ,,	Ellesmere Port	,, ,,
Chepstow	Cartons (using hardboard pulp)	Sittingbourne	Paper

On an outline map of the British Isles, using your atlas index to help you, mark in and name these towns. Use a key which shows their products. Give your map a title.

5. i) *What is meant by forest conservation?*
   ii) *Name two ways in which conservation is carried out.*
   iii) *What is meant by reafforestation?*
   iv) *Name three countries in which reafforestation is in progress.*

5. Minerals

Minerals are one of the natural resources of the world. They exist in various areas through what might be called accidents of nature. Coal, for example, lies where millions of years ago great tropical forests existed, forests which have been buried during past ages and have changed in composition to form coal. Oil, it is thought, formed from the remains of marine life embedded in the floors of ancient seas. Most other substances, such as metallic ores, were formed naturally from the material of the interior of the earth. They often occur in layers, or veins, where particular chemicals have accumulated during the gradual solidification of the earth from its original molten state. The rock is generally known as ore. If the metallic content is high it is called high-grade ore, if low, low-grade ore. High-grade ores are most profitable to mine, but it may also be worthwhile to extract low-grade ores if they exist in large enough quantities.

Study the picture of the open-cast mining. Write a description of it. The Mesabi is a range of low hills in Wisconsin, U.S.A., near Duluth on the shore of Lake Superior, formed almost entirely of iron ore. You can see that deep mining is unnecessary, for the ore is easily accessible from the surface of the land (fig. 1.14).

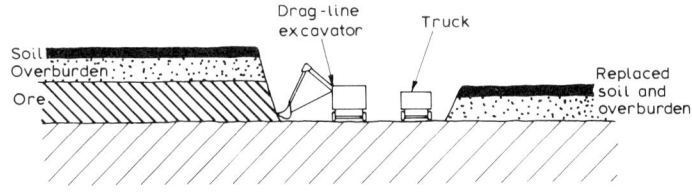

Fig. 1.14. Open-cast working.

Open cast copper mining: Utah.
This is a different type of open-cast mining from fig. 1.14. What are the differences? What mine mentioned in the text looks like this? (*U.S. Information Service*)

When ores occur in this way open-pit or open-cast mining is possible. It is much cheaper than deep or shaft mining (fig. 1.15), and less dangerous. Usually it is only necessary first to remove the soil to expose the ore. The removed soil and over-burden used to be piled in great heaps, and when the ore had been extracted the open pit was just abandoned, usually soon to fill with rain-water. Nowadays in most areas attempts are made to conserve the soil and to restore the land to its original appearance. In Northamp-tonshire, for example, where iron ore lies near the surface, the top-soil is removed very carefully and then replaced, so that the land can be used again for agriculture.

Sometimes open-pit mining is possible even in mountain areas. The Chuquicamata copper mine of North Chile, for example, lies at a height of over 3000 metres in the Atacama Desert, where rainfall averages about 25 mm a year. Here the problem is not the restoration of land for agriculture, since there is none, but that of supplying water and food for the 25 000 people who live in the area. Water is brought from the streams of the snow-capped Andes, and food by lorry and rail from the port of Antofagasta and the oasis town of Calama (fig. 1.16). Chuquicamata is the biggest single copper mine in the world, producing 300 000 tonnes of copper yearly. Of the 17 000 tonnes of copper ore mined daily, some 16 000 tonnes are waste rock. What grade of ore is this? To save burdening the railway with the export of waste rock, smelters have been built by the mine. A smelter is a plant for extracting metal from ore, usually by heating the ore to very high temperatures. The heat here is generated from electricity transmitted from a power station near Antofagasta. The waste from the smelter is piled around in vast slag heaps.

Such waste heaps are familiar sights in our own coal-mining country, and they form another problem of mining. The material here is waste rock excavated from the galleries in deep mines, and brought to the surface mixed with coal. Such heaps are not only unsightly but dangerous; the great disaster in 1966 at Aberfan in South Wales when a 250 metre-high waste tip slipped down and engulfed a village school is probably known to you. Nowadays such waste material is often packed back into the galleries when the coal seams have been removed. This not only prevents the ugly waste heap but helps to support the galleries so that they do not collapse when left. The subsidence of abandoned galleries sometimes affects the land surface, so that houses built on top of underground coalfields may suddenly collapse, or roads subside.

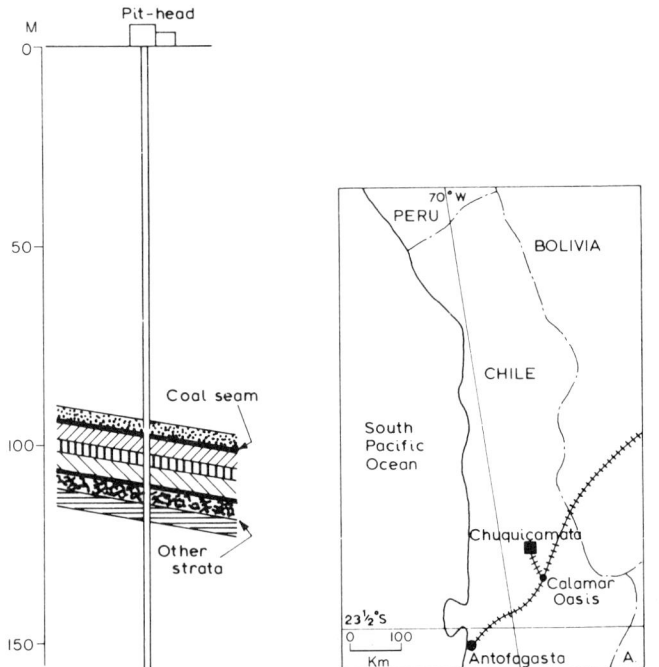

Fig. 1.15. Deep shaft mining.   **Fig. 1.16. Copper mine at Chuquicamata.**

You are all familiar with the other problems of deep mining for coal. Lift shafts are necessary for miners to reach the coal seams, and ventilating shafts so that they are supplied with fresh air. There is danger of fire-damp, a deadly gas given off from coal seams which may cause terrible explosions if ventilation is poor. Pumping is also necessary, for as you know some rocks are saturated with water, which tends to seep out into any holes being dug. The flooding of mines was a great problem before the invention of the steam pump, and even nowadays an unexpected onrush of water may cause drowning.

Another problem of deep mining is heat. As one goes down into the earth temperatures gradually rise. The Witwatersrand goldfield, south of Johannesburg in the Transvaal, is a great gold-field where the gold-bearing rocks, called 'reefs', dip steeply under an area 80 km long from west to east and 30 km from north to south. Here 3600 m deep is the limit for profitable mining, but even at a depth of 2700 m the temperature is 43° C. White South Africans and Europeans cannot do physical work at these temperatures, so labour has been a problem. At one time thousands of Chinese were employed, but now the actual mining is done mainly by Africans. In order to protect them from diseases like phthisis and silicosis, which result from constantly breathing in rock dust, the air in the mines has to be kept saturated with moisture. Gold is not extracted by smelting, but by a process called 'milling'. In 1898 for every tonne of reef which was milled, 14 g of gold was extracted. The ores now milled sometimes yield only 6 g per tonne. In sixty years the Rand, as it is called, has produced 14 million kg of gold. If you like arithmetic you could work out approximately the amount of reef excavated. The waste heaps around Johannesburg are enormous.

In some mining areas the problem is not heat but cold. Two large hills of high-grade iron ore lie near Kiruna in Swedish Lapland. The ore was first obtained by open-cast working in huge terraced quarries, but most areas are now reached by shafts. During the long, cold, dark Arctic winter work progresses only with the aid of powerful arc lamps, and it is difficult to keep machinery working. Workers are paid high wages and given well-equipped and heated houses. Similar conditions are found at the vast open-pit iron mine at Schefferville in Labrador, where July is the only month without frost. At both Kiruna and Schefferville transport was a problem. From Kiruna a railway was built to the ice-free port of Narvik in Norway for use in the winter months; in summer a railway to Lulea, a port on the Baltic Sea, is used (fig. 1.17). From Schefferville the ore goes south on a fifteen-hour rail journey to the port of Seven Islands on the St. Lawrence River.

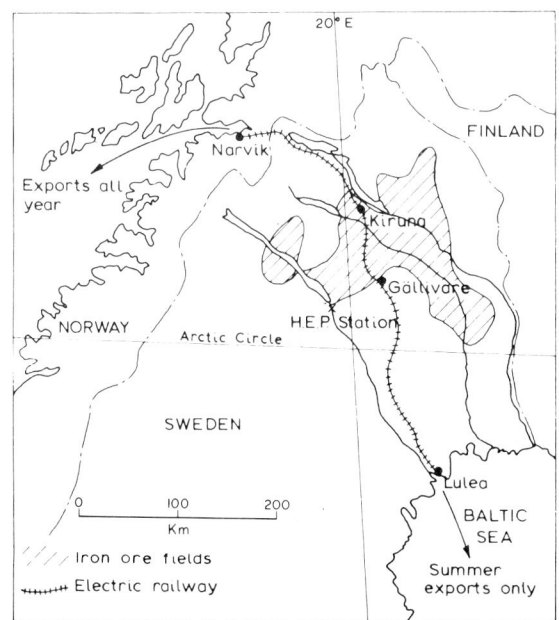

Fig. 1.17. Iron ore fields of northern Sweden.

25

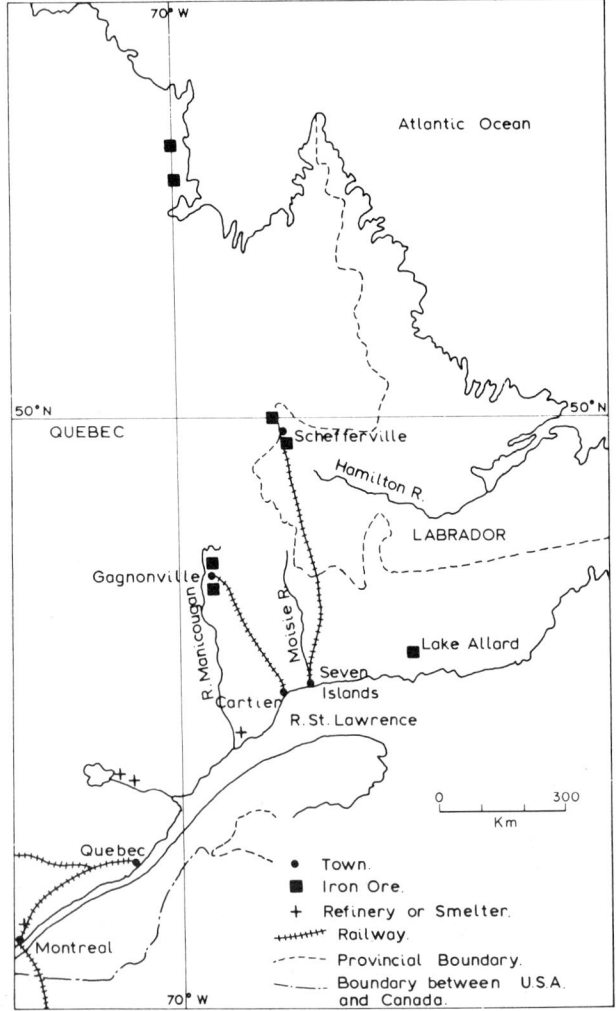

The railway was extremely difficult to build, and is used also to take food and equipment to the miners (fig. 1.18).

Some minerals occur not as ore or in veins of rock but in what are called 'placers'. A placer is a deposit of sand or gravel which contains particles of the required substance in sufficient quantity to be workable at a profit. Gold, tin, platinum and diamonds often come from placers. Early prospectors found gold by hand-sieving or panning placers. Nowadays many placers are worked by hydraulic mining. A jet of water under heavy pressure is directed against the bank of sand or gravel. The material thus loosened is swept by the water into a sluice, where the required particles sink to the bottom, while the sand or gravel is carried away through the sluice. The valuable particles can then be collected up. Much of Malaya's tin is mined in this way, so is the kaolin or china clay of St. Austell in Cornwall. Hydraulic mining is relatively simple; a serious problem which arises is the washing away of soil from entire hillsides and areas of land which might have been useful for agriculture.

Oil, being liquid, is drilled for (page 85). Often much searching is necessary to discover it, and many bores are unsuccessful. In boring the 'bit' which does the actual drilling may become detached and block up the narrow shaft, being difficult to recover. Sometimes a well catches fire. Oil fires are difficult to put out; indeed, there are only a few men expert enough to extinguish the flames by placing a cap over the well to exclude air. Sometimes the well dries up unexpectedly and has to be abandoned. Sea-rigs, used in drilling for under-sea oil, are big platforms on legs anchored to the sea-bed; they may become detached during heavy storms so that the rig sinks. As you probably know from television, some oil companies employ a man called a 'trouble-shooter' to deal with the various problems which arise in oil production.

◀ Fig. 1.18. Iron ore fields of Labrador and Quebec.

Hydraulic tin mining: Nigeria.
(*J. Allan Cash*)

Sea rig in the North Sea.
List the problems facing the managers of this oil rig.

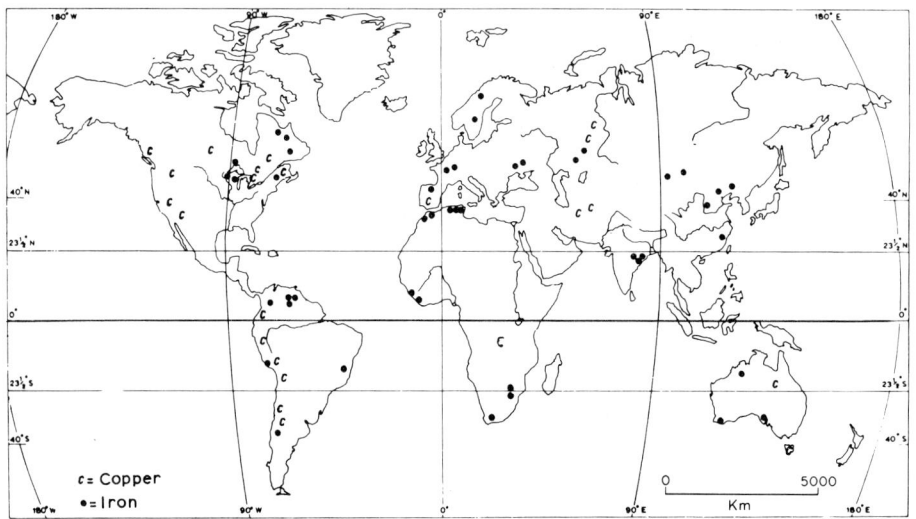

Fig. 1.19. World iron and copper areas.

c = Copper
• = Iron

The map showing the distribution of various minerals (fig. 1.19) will suggest that mining problems exist the world over. They can be summed up as follows. First, the ore must be found. It is then necessary to decide if it is worth the expense of mining. Nowadays large companies usually buy the land containing the ore, or the right to mine it, and provide money or capital for equipment and wages. It may be necessary to build roads and railways to the mines, and houses for the workers. There will also be problems, particularly in deserts or mountains, of supplying food and water. There is the problem of attracting workers to remote mines, or to those which lie in very cold or very hot climates. The methods of mining must be considered. These may involve all the problems of mining at depth, including diseases to which miners are prone. After the ore has been excavated most metals have then to be extracted. If the ore is of low grade it is better to have the processing plant near to the mine. The smelters for the great open-pit copper mine at Butte, in Utah, U.S.A., for example, are located on the outskirts of nearby Salt Lake City. In Africa the copper of the open pits in Katanga is refined with coal sent from the Wankie coalfield of Rhodesia. The aluminium ore or bauxite mined in Jamaica is shipped for smelting via the Panama Canal to Kitimat on the coast of British Columbia, because hydro-electricity is available there from the Kemano power station 80 km away. After extraction, the pure metals have to be sold for use. Iron is so important because it is used in the making of steel. Copper is vital in the transmission of electricity (page 54). Aluminium produces a light, durable metal. Tin is used in the canning industry. Indeed, many minerals are important, and we have discussed but a few.

There are many other problems of mining, not the least being that of the exhaustion of supplies. Mining is a 'robber' industry, that is, it takes from the land natural resources which are impossible to replace. There are many 'ghost' towns, like those of Mammoth and Tiger in the San Pedro Valley of Arizona, which grew up when mining was at its height and were completely abandoned when mining finished. Our own coalfields provide an example of how mines or pits can be closed down, with the resulting problems of unemployment. As coal supplies get used up man turns to other sources of power, such as hydro-electricity and oil. Oil is further discussed in Chapter 4; at one time it was feared that supplies would run out, but fortunately there are still new discoveries in the bed of the Bay of Biscay and the Caspian Sea, in India, Indonesia, Syria, Alaska and elsewhere. New reserves of many other valuable minerals, including gold, coal, copper, iron, bauxite have also been discovered recently. It would seem that man has not yet to face the problem of being without them.

Work to do

1.

i) *The table shows the oil-producing countries of the world. Divide the list into four groups by writing the names of the countries under the following headings:*

a) *North and South America*
b) *Middle East*
c) *Far East*
d) *Other countries*

ii) *Which of the four groups produces the most oil?*

Country	Production in million tonnes	Country	Production million tonnes
Algeria	*47*	*Nigeria*	*54*
Canada	*61*	*Persia (Iran)*	*192*
Egypt	*16*	*Qatar, Oman, etc.*	*100*
Indonesia	*42*	*Saudi Arabia*	*177*
Iraq	*76*	*U.S.A.*	*475*
Kuwait	*137*	*U.S.S.R.*	*353*
Libya	*162*	*Venezuela*	*194*

2. *Describe the differences between open-pit and deep-shaft mining.*

3. *The following table lists the main countries producing iron ore:*

Country	Million tonnes	Country	Million tonnes
U.S.S.R.	*106*	*China*	*22*
U.S.A.	*53*	*Brazil*	*20*
Australia	*33*	*Sweden*	*20*
Canada	*30*	*India*	*19*

i) *Draw column graphs to show the production of iron in these countries, using a scale of 2 cm to 10 million tonnes.*
ii) *Name any iron ore mines you know in any of these countries.*
iii) *Name any uses to which iron is put.*

4. i) *What is mineral oil?*
ii) *Why is it necessary to drill for oil?*
iii) *What problems may arise during oil production?*
iv) *Why are sea rigs sometimes necessary?*

5. *In what ways can climate be a problem in mining?*

6. Earthquakes and volcanoes

If you were asked which country you know to suffer from earthquakes you would no doubt say Japan. Mild earthquakes or earth tremors are so common in Japan that they scarcely make newspaper headlines. Many other countries suffer earthquakes, and the table below lists the major ones experienced from 1964 to 1967.

Look for these countries in your atlas. You will see that they are mountainous countries. All the earthquakes occurred in the mountains near the coast. Earthquakes and mountains are frequently closely related. The mountains were formed millions of years ago when great folds occurred in the earth's crust. Great earth movements slowly heaved up these huge mountain ranges.

Country	Date	Main effects
United States California	Feb. 1971	Caused by movement of San Andreas Fault. 120 people killed by falling masonry.
South America Chile	July 1971	In area near Illapel. 13 000 houses damaged. 100 people killed in Valparaiso.
Peru	June 1968	41 people killed in Amazon Basin region, N. Peru.
Peru	May 1970	Coastal towns destroyed. Severe flooding. 50 000 people dead, 1 million homeless.
Europe France	Aug. 1968	Pyrenean village destroyed. 1100 people homeless.
Greece	May 1967	Widespread damage in Pindus Mountain range.
Greece	Feb. 1968	Two islands in N. Aegean Sea badly damaged.
Italy	Jan. 1968	Many towns destroyed in W. Sicily. 300 people killed, 80 000 homeless.

Country	Date	Main effects
Europe Yugoslavia	Nov. 1967	Border town of Debar in ruins. 7000 homeless.
Yugoslavia	Oct. 1969	Industrial town of Banja Luka damaged.
Asia India	Dec. 1967	172 killed in town 125 km S.E. of Bombay.
Indonesia	Feb. 1967	Devastation in area round Malang, E. Java.
Indonesia	July 1971	Tidal waves wrecked islands of New Britain.
South West Asia Iran	July 1970	175 people killed in border town of N.E.
Iran	Apr. 1972	5000 died in Ghir region of S. Zagros.
Turkey	Mar. 1969	Tremor along Anatolian Fault caused 53 deaths.
Turkey	Mar. 1970	Gediz, S. of Istanbul, destroyed. 1800 people killed, 7000 injured. Great damage.

and the crust of the earth is still unstable or unsteady along the mountain flanks. Present-day earthquakes are short, sharp movements of the earth's crust.

You know that when you throw a stone into a pool or lake a series of waves or ripples spreads through the water in all directions. In the same way, when underground rocks are fractured ripples called vibrations spread out in all directions from the source or centre of the disturbance (fig. 1.20). The passage of these vibrations is the earthquake. The vibrations or tremors can be detected and measured by an instrument known as a seismograph. The intensity or force of earthquakes varies so much that a table, given below, has been made to describe them.

You can now decide which intensity and description should be given to the earthquakes in the table.

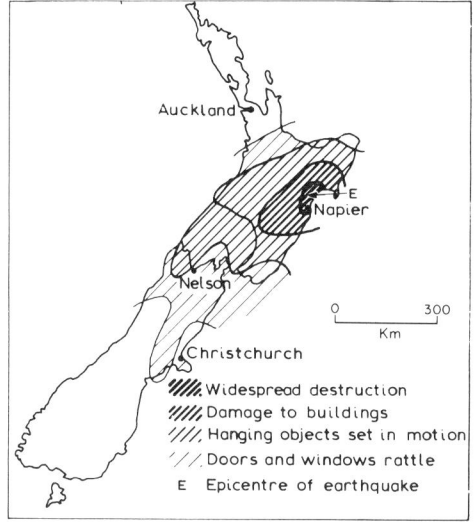

Fig. 1.20. Earthquake disturbance: New Zealand.

	Intensity	Description of effects
1.	Instrumental	So slight that it can be detected only by seismographs.
2.	Feeble	Noticed only by sensitive people.
3.	Slight	Like the vibrations due to a passing lorry.
4.	Moderate	Felt by people while walking; loose objects rock.
5.	Rather strong	Felt generally; most sleepers are awakened.
6.	Strong	Trees sway and all suspended objects swing; damage by overturning and falling of loose objects.
7.	Very strong	General alarm; walls crack; plaster falls.
8.	Destructive	Car drivers seriously disturbed; masonry or buildings crack; chimneys fall; poorly constructed buildings damaged.
9.	Ruinous	Some houses collapse where ground begins to crack; pipes break open.
10.	Disastrous	Ground cracks badly; many buildings destroyed; railway lines bent; landslides on steep slopes.
11.	Very disastrous	Few buildings remain standing; bridges destroyed; pipes, cables, railways destroyed; great landslides and floods.
12.	Catastrophic	Total destruction; ground rises and falls in waves.

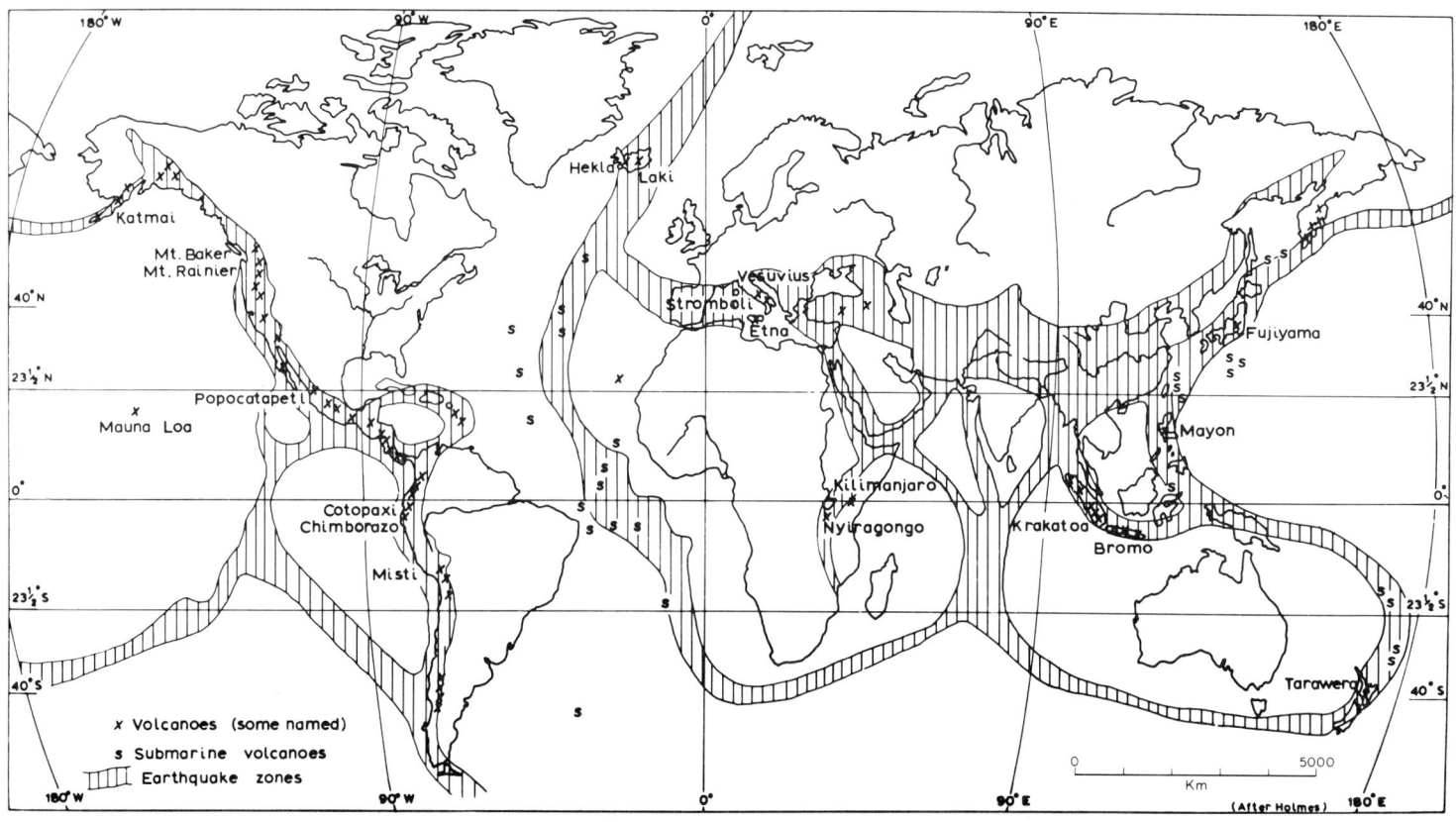

Fig. 1.21. World earthquake zones and volcanoes.

The map (fig. 1.21) shows you the earthquake zones of the world. It also shows you where some of the world's most serious volcanic eruptions have taken place. You will notice that volcanoes, too, are frequently related to or associated with fold mountains; many are on ridges rising from the sea floor or on islands. A volcano is a hole or vent in the earth's crust through which molten rock called lava, hot rocks, mud, steam and various gases are ejected (fig. 1.22). The materials which are ejected or

Volcano in the sea: Fayal, Azores.
Compare this picture with fig. 1.22.
(*Aerofilms*)

Crater within a crater.
All this bare ground is lava.
(*J. Allan Cash*)

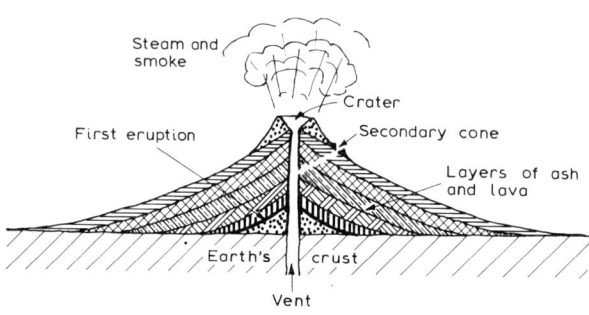

Fig. 1.22. Diagram of a volcano.

thrown out fall around the hole or crater, and gradually build up into a cone-shaped mountain. The great pressures which were exerted when fold mountains were formed are thought to have produced the volcanic vents in weak rocks, so that many volcanoes lie within the belts of fold mountains. Examples are Vesuvius in Italy, Etna in Sicily, Fujiyama in Japan, Chimborazo and Cotopaxi in the Andes of South America and Popocatepetl in Mexico.

When a volcano has not erupted during time known to man it is said to be extinct. If a volcano has not erupted over a long period of years it is said to be dormant (sleeping). Active volcanoes are those known to erupt. The table on page 34 tells you of recently active volcanoes.

Country	Volcano	Date of Eruption	Effects
Chile	Villarica	Mar. 1964	Eruption started a series of avalanches which destroyed a small town 800 km south of Santiago.
Chile	Villarica	Dec. 1971	Lava melted snow, causing flash floods. 30 dead.
Costa Rica	Irazu	Began in 1963, cont. 1964	Covered land with ash for 400 km around; destroyed 10% of coffee crop; 40 000 dairy cattle moved to safety.
Indonesia (Flores I.)	Mt Ija	June 1969	100 000 people homeless. 1300 hectares of plantations destroyed.
Italy (Sicily)	Mt Etna	April 1971 to June 1971	Worst eruption since 1928. Lava diverted to save towns of Fornazzo and Sant Alfio. 200 hectares vines, orchards, chestnut groves destroyed. A wide belt of rich volcanic soil lost to agriculture.
Sangihe Island (Indonesia)	Mount Awu	Aug. 1966	Eruption killed 15 people, injured 1100, and made 21 000 islanders homeless.
		Sept. 1966	Second eruption made 14 000 people homeless.
U.S.S.R. Azerbaijan	A new volcano	Jan 1965	New volcano suddenly erupted in an oilfield.

When a volcano is about to erupt the ground around it usually shakes and trembles, producing an earthquake. Then steam and gases rise from the vent or crater. The steam gives rise to the formation of clouds over the volcano, and often heavy rainstorms follow. Finally, there is often an explosion, and heated rocks and ash are hurled into the air. Lava pours from the crater and flows down the mountain sides, destroying all in its path. In a violent explosion the whole cone round the crater may be blown off. Some volcanic eruptions are not preceded by explosions; the lava wells quietly out of the crater and flows down the mountain sides until it gradually cools and hardens.

Volcanoes and earthquakes are natural hazards. Earthquakes are particularly disastrous to man because he does not know when or where they will occur. They may cause whole towns and villages to collapse. Bridges fall down, railway tracks are torn up or bent. Water-pipes burst, gas mains and electricity cables are broken. Fires often break out, and are difficult to put out because there is no water supply. Many people may be killed or injured. Severe earthquakes may cause rivers to leave their beds and flood the surrounding land. Dams may burst. Earthquakes in islands or on the sea-bed disturb the sea so that huge tidal waves flood over neighbouring coasts, drowning many people and destroying harbours and ships. In all cases rescue work is difficult because roads and railways may have been destroyed. The destruction of sewage systems may cause diseases to break out, often drinking water is contaminated and typhoid occurs (page 40). The dangers of volcanic eruption can sometimes be avoided, because the volcano may show signs of activity for several days or even weeks before it erupts seriously, and people living near by may be able to move away. Nevertheless, violent volcanic eruptions can destroy whole towns, burying them in lava, dust or ashes. In 1883 the explosion of Krakatoa, a volcano on a small Indonesian island of the same name, was heard in Australia over 3000 km away. It blew much of the island into the ocean, and created huge waves which caused the sea to rise 15 metres on the shores of neighbouring islands so that 36 000 peóple were killed.

When natural disasters such as earthquakes or great volcanic eruptions occur the nations of the world all co-operate to help. They send tents to house the homeless; blankets, clothes, food, medical supplies, hospital equipment and transport. Doctors and nurses volunteer to help, teams of workers and soldiers from nearby countries are rushed out to aid in rescue work, clearing away debris, restoring water supplies and communications. Countries raise relief funds of money to help the rebuilding of destroyed towns and villages, and for people who have lost all their possessions. Man can do nothing to prevent these natural disasters, although nowadays in some countries observatories are built near volcanoes which are likely to be active so that scientists can watch them and forecast likely eruptions. It may then be possible to evacuate the people living nearby before disaster overtakes them.

Work to do

1. *Write a description of the damage shown in the picture caused by the earthquake.*

2. i) *Name three countries in which earthquakes occur.*
  ii) *Write a brief explanation of what is meant by an earthquake.*
 iii) *Describe the intensity or force of an earthquake called 'disastrous'.*
 iv) *Why is rescue work difficult after a very disastrous earthquake?*

Earthquake damage: southern Chile. ▶
(*Camera Press*)

3. 'Thousands of people were feared to have died in an earthquake which rocked Eastern Iran yesterday. About 2000 square km in the mountainous east and north-east of the country were shaken by the tremor which lasted between one and three minutes. Worst hit was the area around the village of Kakhak, in the centre of Khorassan province. Many thousands were made homeless by the earthquake, which shook more than a hundred towns and villages. They camped out in the open air, waiting for tents, blankets, food and medical supplies to be airlifted into the stricken area' (Newspaper report, 1 September 1968).

   i) Find Iran in your atlas and locate the earthquake area.
   ii) Why do you think so many people died or were rendered homeless?
   iii) How many towns and villages were affected?
   iv) Why were tents, blankets, food, etc., to be airlifted into the area?

4. Study the map (fig. 1.21) and answer the questions.

   i) Using your atlas, name the main mountain ranges in the earthquake areas.
   ii) Write two sentences describing the position of the earthquake areas.
   iii) Name three volcanoes lying in mountain areas.
   iv) Using the text to help you, explain why earthquakes and volcanoes occur where they do.

5.   i) Copy the diagram of the volcano, labelling it carefully.
    ii) Write a sentence on each feature of the volcano, explaining how it is formed.

Floods near Aylesbury. ▶
(Aerofilms)

Chapter 2

PROBLEMS OF WATER

1. Water supply

Man, animals and plants cannot live without water. In our country we are so used to merely turning a tap to get unlimited supplies of this precious liquid that we seldom think how fortunate we are compared with many other countries where water is in short supply, nor do we stop to think of the problems involved in supplying us with water. In many towns in England water is used at the rate of over 200 litres per person per day. Where does it all come from?

Basically the water that we use comes from our rainfall. Some of this soaks into the ground. If the soil is underlain by permeable or porous rock, as in the diagram (fig. 2.1), the water collects in it. Porous rock may be made up of tiny particles with equally tiny spaces in between them, and in these spaces the water collects. In the diagram the porous rock, in this case the chalk of the North Downs and Chiltern Hills, collects water. The chalk is underlain by an impermeable or non-porous rock, clay, which prevents the water from seeping any farther. If a well is bored into the porous rock the underlying water either gushes out or can be pumped up. Such a well is called an artesian well (fig. 2.2), and there are hundreds of such wells in London, some of them 200 m deep.

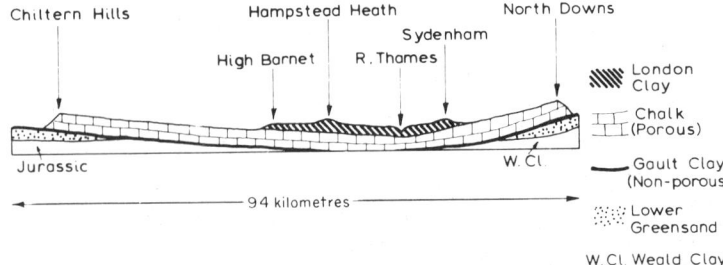

Fig. 2.1. The London artesian basin.

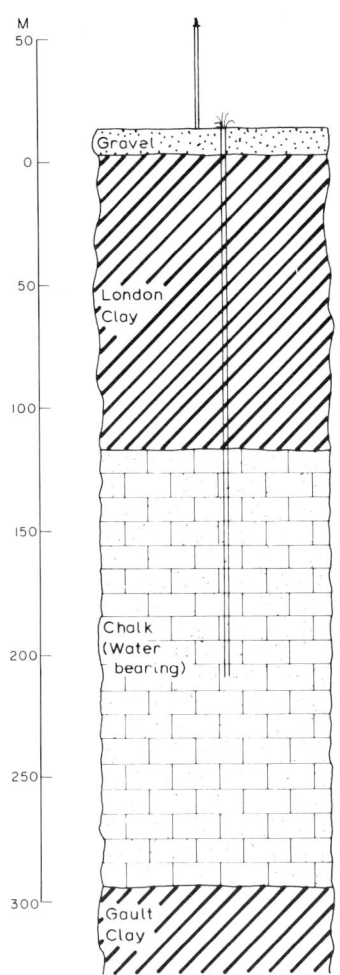

Fig. 2.2. Artesian well: Trafalgar Square.

Edinburgh's water supply: Talla Reservoir.
Note the absence of population in the gathering ground of this reservoir. The structure in the reservoir is the draw-off chamber, or valve tower, which gives access to the controls on the outlet pipes for the water.
(*Camera Press*)

The fountains in Trafalgar Square and the roof garden on Derry and Toms' store in Kensington High Street are supplied with water in this way. Many factories in London use artesian water; indeed, so much water has been extracted from the chalk that new wells are not now permitted for them.

In many parts of Britain the rock underneath the soil is non-porous, so that the surface water does not sink away, but collects on the surface in streams, rivers, ponds and lakes. These form the greatest source of water which we use, but the water has to be purified or cleaned first. About 85 per cent of London's water is pumped from the River Thames, on the banks of which it is purified from large reservoirs. As the population of London has increased, it has been necessary to provide more reservoirs. The Chingford reservoir in the valley of the River Lea, a tributary of the River Thames, was completed in 1913, and in 1925 another large reservoir was built near Staines. There is now a plan to use the underlying chalk itself as a reservoir.

The heaviest rainfall we receive falls in the highlands of our country, and these are the sources of most of our rivers. They are also the home of most of our lakes. The water supply of people living in the Scottish Highlands, the Lake District, the Pennines and Wales presents no problems, but these are the areas where fewest people live. It has therefore been possible to take water from these areas through pipes to the big industrial cities where it is needed. Lake Vyrnwy in the Berwyn Mountains of Wales is now a reservoir which supplies Liverpool, and Birmingham

Water supply protection: Loch Katrine.
Draw a sketch map to locate the places named.
(*L. J. Long*)

LOCH KATRINE
GLASGOW'S WATER SUPPLY.
YOUR CO-OPERATION IN KEEPING THIS SUPPLY PURE IS REQUESTED.
CAMPING, BATHING, PADDLING, FIRES, PICNICKING AND LITTER ARE STRICTLY PROHIBITED.
FISHING FROM THE SHORE NOT PERMITTED.
PLEASE KEEP TO THE ROAD AND DO NOT THROW COINS INTO THE LOCH. THANK YOU.
ENGINEER AND MANAGER,
LOWER CLYDE WATER BOARD,
50 JOHN STREET, GLASGOW, C.I.

receives water from three reservoirs near Rhayader in the valley of the River Elan, a tributary of the upper Wye River, in central Wales. Lake Thirlmere in the Lake District supplies Manchester, and several rivers in the Pennines have been dammed to create reservoirs for Sheffield.

We must also remember that water is not only necessary for use in homes but is essential for industry. The soft water of some of the Pennine streams, such as the River Ribble in Lancashire, has aided the development of the cotton industry, for it has long been used in washing the raw cotton and in dyeing the woven fabrics. Similarly, the waters of the Rivers Aire and Calder are used in the woollen industry. Brewing beer uses vast quantities of water in places like Burton-upon-Trent, and the paper mills of the Thames estuary need a steady supply. Many industrial plants need a supply of water to cool their engines, just as a radiator is used to keep a car engine cool. The water used by industry is in itself a problem, as it may flow away into rivers or streams and contaminate them, that is make their water impure (page 82).

Although most people in our country have a water supply piped to their homes, those in remote and isolated houses may still have to pump their own water from a well or even carry it from a distant spring (fig. 2.3). Where the ground is solid rock, as for example in the Isles of Scilly, rain-water is collected from the roofs and run into a tank. The collection of rain-water in this way occurs in many parts of the world, and in the island of Bermuda government law insists that roofs be repainted white each year so that the rain-water is kept as clean as possible.

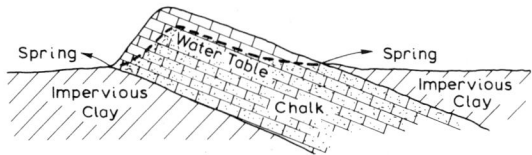

Fig. 2.3. Water-table and springs.

The provision of water for use in homes is a great problem. In the Ganges Valley of India water is drawn from wells dug in the alluvial soils; these wells provide water even during the dry season. In the eastern Deccan of India many of the streams flow only during the wet season, so small dams are constructed across them, behind which the water collects in ponds called tanks. These ponds are used for all purposes. Pots are dipped into them to collect water for drinking and washing, cattle drink at them, and water is channelled from them to irrigate the land for crop production. The water is used carefully because it has to last until the rains come again. If it is in short supply the irrigation water has to be cut off, then crops fail and famine may result.

In mountainous countries the water supply is usually constant, because rainfall is heavier and snows on the higher peaks melt in the spring and summer, so adding to the water supplies. Nevertheless, many of the people who live, for example, in the Himalayas may have to walk some distance to the nearest stream. In one remote village in Nepal some British explorers recently presented the villagers with a kilometre of plastic pipe, like a garden hose, the end of which was placed firmly in the mountain stream. The water supply from this pipe, which led into the village square, saved the village women several hours walk daily, and saved them the burden of carrying water back from the stream three times a day.

In many countries of Asia and Africa water is scooped up directly from the rivers. This work is often done by the women, who carry the heavy vessels of water back to the little farms and villages. The river water is often very impure, although the people have become generally immune to the diseases it carries. Bilharzia is a disease which causes ill-health rather than death in much of Africa. The germ lives in a small water-snail, and enters the human body through the skin of feet or hands when washing or bathing in rivers. It is particularly common in the irrigation canals of Egypt. Sometimes such impure water gives rise to outbreaks of typhoid, a fever which spreads rapidly and causes many deaths. Outbreaks of typhoid have occurred even in our own country and in Switzerland when normally purified water has become contaminated by accident. Visitors to countries using unpurified water are usually advised to boil all water before using it. The cost of purifying water is high, and we help to pay for our clean water when we pay water rates, collected by our Water Boards twice yearly. There are millions of people in Asia and Africa who at present do not have the money to pay for purified water if it were available.

In lands of limited rainfall the problem of water supply is indeed difficult. In deserts with a rainfall of less than 250 mm a year, and sometimes no rainfall at all, there are few people. In the Sahara water occurs at oases. This water has seeped underground from rainy areas surrounding the Sahara hundreds of km away, and oases develop where it is sufficiently near the surface for seepage into wells or for tapping by the long-rooted date palm (fig. 2.4). Because oases provide water they are generally the only

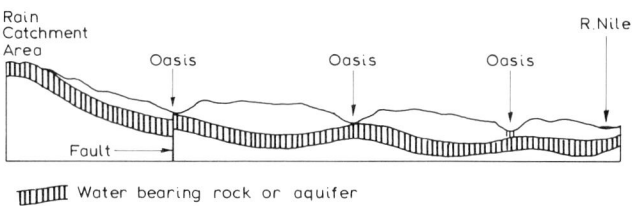

Fig. 2.4. **Underground water in the Sahara.**

areas in deserts where people can settle to live. Oases occur in the Atacama Desert of South America also. When valuable minerals are found in desert areas it may be necessary to carry water to the mines in tanker lorries, or it may be worth while, as in the case of the gold mines of Kalgoorlie and Meekatharra in Western Australia, to lay water pipes from the rainier coastlands.

The provision of fresh water is almost certainly the most important technical problem of our time. To provide it cheaply presents the greatest challenge to the world's scientists and engineers. Since the fresh-water resources of the land are likely to prove inadequate for the ever-increasing world populations, scientists are turning their attention to the sea. In 1965 an international conference was held in Washington, U.S.A., the subject under discussion being the economic production of fresh water from sea-water. It was attended by 2000 delegates from sixty-five countries, so that it is clear that the problem is of international interest.

Sea-water contains about 35 000 parts per million of dissolved salts; drinking water should contain less than 500 parts per million. From a purely scientific point of view, the production of fresh water from brine is not difficult. There are many ways of producing it, the two most easy to understand being distillation and freezing. Distillation depends on the fact that steam produced when sea-water is boiled is free from salt. Freezing makes use of the fact that when sea-water freezes the ice formed is pure water. It is likely that people living in the Arctic had observed a long time ago that ice formed from sea-water could be melted to produce fresh water. Whatever process is used, heat is needed, and if the fresh water is to be produced on a large scale expensive equipment is necessary. The machinery for making salt water into fresh water is called a desalination plant.

The opening up of new water resources often involves high costs for building dams, reservoirs and pipelines. These costs might well be higher than those for setting up a desalination plant. Such plants would need to be placed where there is sea-water and cheap power available. They should be placed in areas where they would not make the countryside unsightly. In England and Wales particularly the areas of high rainfall cover most of the national parks, where people are reluctant to build dams and reservoirs which spoil the countryside. The cheapest form of power in our country may one day be the electricity produced from nuclear reactors, and there are experiments going on to see if it is possible to produce desalinated water at one of these nuclear power stations at Troon on the west coast of Scotland.

Although these plants are still in the experimental stage for Britain, British desalination plants have been in operation in some countries for the past ten years or more. The location and capacities (in million litres per day) of some of these plants is tabled below:

Kuwait State .... 36·0
Curacao .... 27·0
Qatar .... 9·0
Kuwait Oil Company .... 4·5

The last name on this list gives you the clue that most of these plants depend on oil for their power, and they were built because oil was available in vast quantities while fresh water was scarce and the sea was near. There are many other plants scattered round the world in South America, Africa and Australia. The cost of producing this water is far too high for it to be used for irrigation purposes, but when cheaper methods for the desalination of the unlimited supplies of sea-water are available they may solve a great problem of the desert areas of the world today.

Work to do

1. *Copy the diagram* (fig. 2.1). *Label it carefully. Then write a paragraph explaining it.*

2. *Draw a large map of the Lake District. Mark in and name:* (a) *the mountains;* (b) *the lakes and main rivers;* (c) *put an arrow from Lake Thirlmere towards Manchester* (*which will not be shown on your map*).
   Give your map a title.

3. *Write a description of the picture on page 39.*

4. i) *What is a desalination plant?*
    ii) *Name three countries in which such plants are used.*
    iii) *Why is their use at present limited?*
    iv) *In which areas would desalination plants be particularly useful?*

5. *Write a paragraph each about the following topics:* (a) *Artesian water supplies;* (b) *London's water supply;* (c) *water supply in mountainous countries.*

2. Drainage

You have seen that water is vital to our life, and in desert lands provision of water is probably the very first thing that must be considered. But we can also have too much water, and this is broadly the problem of drainage, or how to take surplus water off the land. Fortunately nature is always helping us, and the great majority of the earth's surface is drained by the simple process of water running away downhill, though as you have learnt, much soaks into the ground first, and in warm weather much evaporates. Sometimes when looking at a map pupils see many blue streams marked and say 'the land is well drained', or they may see no streams and say 'it is badly drained'. This is not correct. There could be no streams yet very good drainage if the land were porous and all the rain-water had soaked in. There could be many streams which did not do their job well, or flooded frequently, and this would be bad drainage. So we mean by good drainage a good drainage system, which takes away all the surplus rainfall which man does not want.

There are many signs in nature of bad drainage, apart from the obvious ones of wet, soggy land, or dampness or fungus in the basement of a house. The common reed or rush is one of the first plants to appear when drainage is poor. On damp pastures or moorlands, hardly swampy enough to be called true marshland, cotton grass is commonly seen. This, as its name suggests, has a flower which looks like a ball of cotton-wool, and is easily recognised. When drainage is so bad that the whole area is sodden and squelchy we use the name marsh, fen or bog, and true marsh plants, such as sphagnum moss, are the only ones which can survive. These areas, if they continue to exist for centuries, become areas of peat. Peat is simply an accumulation of vegetation which has not rotted away. In cool damp conditions the bacteria which cause decay cannot live, and leaves, stems and roots just sink down, making a layer often many feet deep. In the

British Isles, central Ireland and parts of the Scottish Highlands have the largest areas of peat, and when they have been drained the peat can be cut and dried for use as fuel. Also on mud-flats by the sea, plants and grasses which can tolerate salt sometimes take root; these areas gradually become salt-marsh, and are sometimes reclaimed to make new land.

You probably think of marshland or badly drained land as a very flat area where water does not run away easily, such as a river delta or other flat land by a stream or by the sea. These

Peat cutting in northern Ireland.
Draw an annotated diagram of this picture. Note the moorland and the depth of the peat. What must be done to the cut peats before they can be used?
(*J. Allan Cash*)

certainly are lands which often need artificial drainage, but there are others, particularly on rainy uplands. If the land is not porous, and fairly level, and a river system has not yet fully developed, large areas of upland bog can be formed. Much of the flat-topped hills of the Pennines are this kind of country.

Now let us see how nature drains the land. This is a long story not yet completely known. To study it is rather like coming in halfway through a film. We can look at the state of rivers now, but they have been developing since the earth began to take shape, and the start of this film was the beginning of the earth—whenever that was. We can, however, observe certain things happening, which may explain the growth of the pattern of drainage. On hillsides there are usually many small streams. These streams gradually cut down into the hillside and make a valley. Down the sides of this valley in turn more small streams flow, forming tributaries. And so on and on. This makes the familiar network of streams which you see on a map, looking like the trunk and branches of a tree. You will often see a pattern like this develop on mud-flats or a wet beach at low tide. The whole area drained by a river is called its basin, and the line along the higher ground separating this basin from the next is the watershed (fig. 2.5).

This is the ideal situation. It does not always happen, and then we find badly drained land. During the glacial period great ice-sheets moved across many lands, scraping away sometimes the established pattern of the rivers, so that the surface was uneven. Here were left many small lakes or flat marshy areas which are not yet joined up again by a continuous river system to take the water away. Also the land is not necessarily shaped into a neat pattern of hills and valleys. There are folds and dips in the ground where water collects, and these remain marshy or ill-drained until the water fills them up and finally overflows at the lowest point on the rim of the depression. It then gradually cuts its way down and drains the area.

Lakes have already been mentioned in this chapter. They are important reservoirs of water. They are also only temporary fea-tures of the earth's surface, if we think in terms of the millions of years of its history. Most big lakes were caused by some blocking of the normal pattern of drainage, often during the Ice Age. The rivers which drain them are constantly cutting through the barrier. The Vale of Pickering in Yorkshire is such an old lake bed.

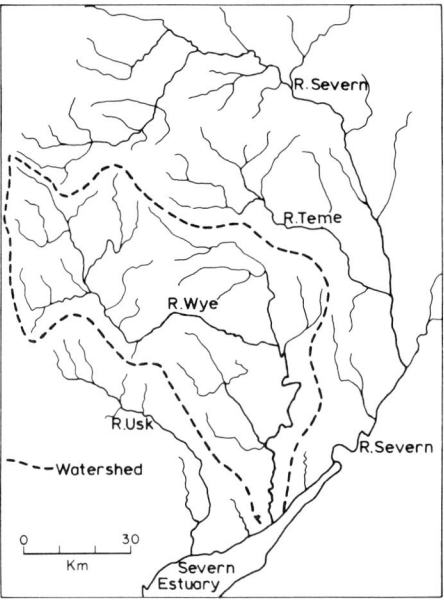

Fig. 2.5. The River Wye basin.

The River Derwent has now cut its way through the Howardian Hills, and the lake has thus been emptied. As its floor is extremely level, owing to the layers of sediment which settled down on it, it is rather marshy, and man has had to assist nature in this case by digging artificial drains across the flat lands (fig. 2.6).

We mentioned marshy land by rivers. Such lands are the main

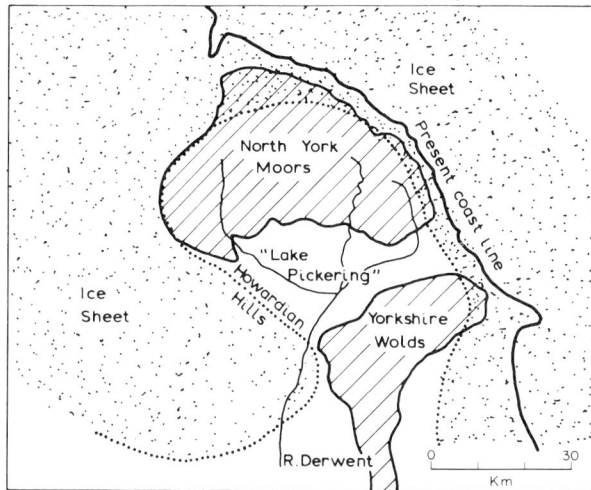

Fig. 2.6. The formation of Lake Pickering.

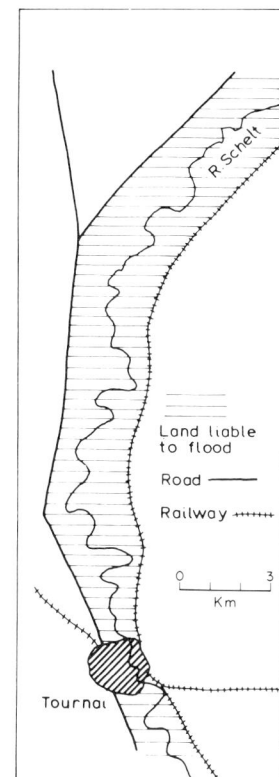

Fig. 2.7. The flood plain of the River Schelt.

regions which need drainage and protection from floods. Let us see how they occur. It is clear that as rivers get bigger on their way to the sea they can erode larger valleys. They also swing about in great curves called meanders. The real cause of meandering is not fully known, but that is how rivers behave (fig. 2.7). The level land across which the rivers swing is the flood-plain because, as the picture on page 37 shows, this is the area which is flooded when the river rises above normal level. Flood-plains are important to man, because they are usually covered with good alluvial soil. But they are also liable to flood, and therefore people in general try to avoid living on them. All rivers have a flood plain somewhere on their course, and large rivers, such as the Nile or the Mississippi, have flood-plains many kilometres wide, and hundreds of kilometres long. Now study exercise 6.

The problem of flood control is obviously a big one. In countries with rainfall which comes all at one season rivers flood so regularly that everyone knows it would be madness to settle on the flood-plain area. But this is not always the case, and the really disastrous floods are those which occur from unusually sudden and heavy rainfall. Advanced countries have river boards, whose task is to watch the level of the river, issue flood warnings, improve navigation and build up the banks. It is unwise for man to interfere with nature too much in this way. The flood-plain is nature's safety valve in time of flood, and by great rivers certain areas must be left unoccupied to allow for this (fig. 2.8).

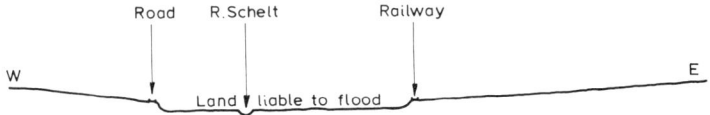

Fig. 2.8. Section across the flood plain.

Most farm land in Britain is drained naturally, but farmers sometimes help by digging field drains. These are simply drain-pipes set in trenches dug in a fish-skeleton pattern, and filled in. These lead down-slope to the ditch which surrounds many fields. If you look closely at them you will see these ditches lead to bigger ones, which in turn become streams. On flood-plains quite large drainage ditches must be dug and kept clear. They must also

Floods in Italy.
(*Camera Press*)

have a sluice gate which can be opened for drainage when the river is low.

It is a short step from the drainage of fields to large reclamation schemes. The best known areas of artificial drainage in England are Romney Marsh, Sedgemoor in Somerset and the Fens. The last named were a large area of peat marsh formed around the rivers running into the Wash. Attempts have been made to drain them since Roman times, but the chief works were by a Dutch engineer named Vermuyden. The whole area is now artificially drained. The new 'rivers' run between high artificial banks, which are wide apart to allow spare room in flood time. They are above

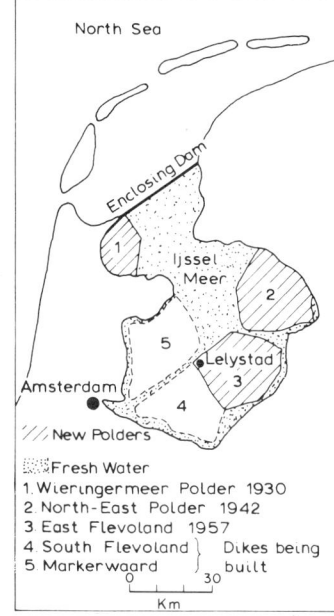

Fig. 2.9. Reclamation of the Zuider Zee.

the level of the rest of the land, and water from the drains must be pumped up to them at pumping stations. All this work was well worth while, for the Fens are now the richest agricultural land in the country.

An even larger piece of reclamation by drainage is being done by the Dutch, in the area formerly the Zuider Zee (fig. 2.9). The first task was to build the great enclosing dam, to keep the tides out, and the first small polder, as the drained areas are called, was finished in 1930. Next a dam was built around the North-East polder. It was pumped dry, and completed in 1942. Some years had to pass before the salt was washed out of the ground, but now it is prosperous farmland, with many villages and a new town. The other polders are still in the course of construction. Eastern Flevoland is almost complete, and the main dikes for the others are beginning to appear.

It is tempting to imagine a similar piece of reclamation by a main dam across the Wash, but the water is deeper and the currents stronger. A further problem is that special locks would have to be built to keep open the small port of King's Lynn, and local residents would certainly object to having their seashore disappear.

Work to do

1. *Trace from your atlas or other map the largest river system in your area, and mark in the watershed. Add a title 'The basin of the River—'.*

2. *Write a description or draw a sketch map of any river flood-plain known to you.*

3. i) *Copy the map of the Zuider Zee area (fig. 2.9), then answer the questions.*
   ii) *For what purpose were the dams built?*
   iii) *What is a polder?*
   iv) *How was the sea-water taken from the polder?*
   v) *Name one large polder area.*

vi) *Is the water-table in the polders near to the surface or well below ground?*
vii) *What are polders used for?*

4. *On an outline map of England mark and name all the areas which need artificial drainage.*

5. *'The farms here are smaller than farms in the nearby uplands and the rest of East Anglia. In Holland only 9 estates exceed 400 hectares, and only 5 in the Isle of Ely. Here many farms are of less than 6 hectares. Although small, the incomes they provide are higher than those from upland farms. The crops usually grown—corn, particularly barley, potatoes, sugar-beet and vegetables—show far higher profits than grass. Animal products are consequently of relatively little importance. In the Isle of Ely there is an average of only 2 cows per farm, and 3 or 4 in Holland, as compared with between 8 and 20 in East Anglia. Sheep are of still less importance.'*

i) *What part of England is here described?*
ii) *How does the size of farm compare with that of the rest of East Anglia?*
iii) *What are the chief crops grown?*
iv) *What is barley used for?*
v) *Why is sugar beet a useful crop?*
vi) *Why are few animals kept here?*
vii) *How can you tell from the text that this area is low-lying?*

6. *'In August and September 1971 excessive monsoon rains caused extensive flooding in much of the lower Ganges Valley, and large areas of Bangladesh were affected. Many towns and villages were under water, and people tried to reach higher ground or fled as refugees to India. Food was scarce and despite world help more people died from starvation than drowning.'*

i) *Find the places named in your atlas.*
ii) *Write a description of what it must have been like in the floods.*
iii) *Write a paragraph to explain why it was almost impossible to get food to the flooded areas of Bangladesh.*
iv) *Draw a sketch map to show the River Ganges and its flood plain.*

3. Irrigation

In Section 1 we considered water supply mainly for drinking and for industry, but we also learnt that plants need water in order to grow. Even in this country, when the natural rainfall is insufficient man tries to add water to the land in various ways. You know how you water your garden by means of a watering can or hose. When you do this you are irrigating. Irrigation is the process of artificially giving water to the land. You will have seen sprinklers turning round to spray lawns: the sprinkler is a mechanical aid for irrigation. You may have seen this on a larger scale. Some farmers have a long row of sprinklers spraying a specially valuable crop, such as vegetables or fruit. The sprinklers are arranged to spray all one row, and then the whole apparatus can be moved to water the next. In England the annual rainfall is generally enough for plant growth, but in a dry summer some find it worth while to give extra water in this way. In Hampshire river water is led out over water-meadows, and this gives a richer growth of grass, even in normal weather.

In general, you will think of deserts as places where irrigation is needed. Cultivation is possible only where a great river provides a water supply, or where an oasis has underground water which can be tapped by a well. But there are other climates where irrigation is useful. Wherever there is a marked dry season, and sufficiently warm weather for plants to grow, irrigation can produce more crops. The Mediterranean lands with their hot dry summers, and the monsoon lands, with their dry but often quite hot 'winter' season, have much irrigation. In general, the tropics and sub-tropics have most irrigation. You should remember also that on the whole irrigation is justified only where cultivation is intensive, that is, where people do a great deal of work on a small amount of ground. It does not pay to irrigate the great wheatlands, for example.

Man has used irrigation for thousands of years, and there are still many parts of the world where the methods used were invented long ago. These methods are simple and cost little for equipment, but they involve long hours of monotonous toil for man himself. The water usually comes from nearby rivers, but nowadays is sometimes brought by canals. The fundamental problem is that of lifting the water up so that it can run along the ditches or channels dug to lead it on to the plots of land.

One method of raising water is by means of what is called in Egypt the shaduf, though the device is found in many other lands. It consists of a wooden pole acting as a lever, supported on a post. One end of the lever has a bowl attached to it, the other end is weighted with a lump of clay. The operator stands on the river bank and pulls the bowl down into the water. When it is full, and thus heavy, he pulls it up, aided by the clay counterweight. He tips the water into a ditch which runs from the riverside to his plot of land, where it runs out among the plants. He has to work hard for many hours to get enough water for his crops.

Another simple method is by the sakia, also known as the Persian wheel. This consists of a wheel fixed at the bank of a river so that some of the rim is dipping in the water. There are water pots attached round the rim. This wheel is geared to a horizontal wheel set on the flat land of the river bank. The farmer harnesses an ox or donkey to the horizontal wheel, and as the animal walks round and round this wheel revolves, and thus also turns the water wheel in the river. The pots each in turn dip into the water, and when they revolve to the top they empty into a wooden channel which leads to the ditch and the fields. The animal spends its whole day doing this, and is often blind-folded so that it does not get giddy. The sakia is used in many countries, including Egypt, India, China and even the European countries of Spain and Portugal.

There are various other devices for lifting water. In China a simple treadmill is used to turn a chain of buckets, and the Archimedian screw is rather like a large corkscrew which fits into a cylinder, dipping in the water. When turned, it forces water up the cylinder and out into the ditch. All these need hard work, and today small mechanical pumps powered by oil or electricity are available. But much irrigation is in the under-developed countries of Africa and Asia, where farmers cannot afford to buy them yet.

You will have realised by now that flood-plains are the most suitable parts of the earth for irrigation, and these are actually where most is found. The land is level at right-angles to the course of the river, but it slopes very gradually downwards in the same direction as the river, otherwise there would be no flow at all. Large modern schemes make use of this fact. A barrage is built across the river, to store water and to raise its level. A main canal is then built running gently downhill along the flood-plain, with side canals taking off at intervals to the fields. Thus the whole area is supplied with water by gravity, and the laborious lifting work is not necessary (fig. 2.10).

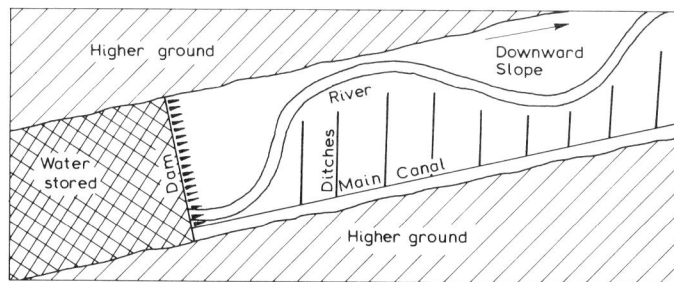

Fig. 2.10. Gravity flow irrigation.

Modern schemes are usually planned by the Government, as they cost many millions of pounds, and are a permanent national asset. They can only be developed where there is a large water

Irrigation ditches: Idaho.
Using this picture and the next, draw your own diagram to show how the water is distributed.
(*U.S. Information Service*)

supply available. One such scheme, which irrigates the semi-desert clay plains of the Gezira in the Sudan, was planned jointly by the British and Egyptian Governments. The map (fig. 2.11)

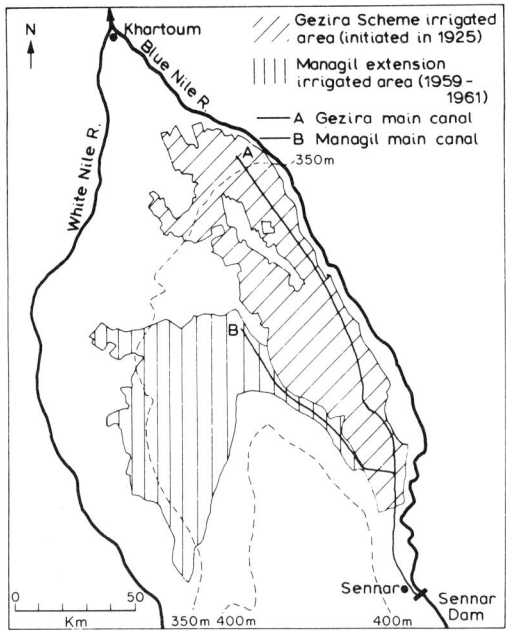

Fig. 2.11. The Gezireh irrigation scheme.

shows you that the waters of the Blue Nile are used to supply two large canals from which many other channels run, though these are not shown. Although the rainfall of this area is only about 250 mm annually, the Blue Nile has ample water because its source is in the mountains of Ethiopia, which receive heavy summer rains. Before irrigation the Sudanese people living in the area could grow only millet, a small-grained cereal which you know as

Distributing irrigation water.
(*J. Allan Cash*)

bird-seed for canaries or budgerigars. In 1925 a great dam was built across the Blue Nile at Sennar: behind it the waters of the river formed a huge reservoir. This reservoir feeds the two main canals: from these the water flows over the whole area shaded on the map. The farmers can now grow cotton, millet and fodder crops for their oxen and other animals in a three-year rotation. They sell the cotton as a cash crop. This helps to pay for their water, and to buy other food and necessities.

The Central Valley Project of California (fig. 2.12) is an example of large-scale irrigation in an area of Mediterranean climate. The Central Valley is drained by two main rivers, the Sacramento in the north and the San Joaquin in the south. They join to reach the sea at San Francisco Bay. The Sacramento has most water, as it is in the wetter north, but this is needed in the valley of the San Joaquin, where the dry summer makes irrigation essential. The Project was therefore designed to transfer the surplus water from the Sacramento to the San Joaquin valley by means of the Delta–Mendota canal. It has to be pumped upwards in places, against the downward slope of the land. There is a network of smaller canals and pumping stations to distribute the water, which is regulated in the first instance at the Shasta dam.

An example of irrigation in a different type of area is in the Murray–Darling Basin of Australia. As you will read (page 202), this is a region of warm temperate grassland, suitable in its natural

Fig. 2.12. The Central Valley Project: California.

Storage reservoir and dam: Lake Eucumbene, New South Wales.
Write notes on the situation of this dam.
(*Australian Information Bureau*)

state only for sheep-grazing and wheat-growing. All the elements necessary for successful irrigation exist (fig. 2.13). There is a rim of well-watered highlands, a river network which flows over level plains, irregular and insufficient rainfall, and high temperatures. The main dams are the Burrinjuck on the Murrumbidgee, the Eildon on the Goulburn and the Hume on the Murray, where the great Snowy Mountains scheme provides most of the water. Large irrigated areas now exist by each river, particularly by Narrandera, Shepparton, Mildura and Renmark. Vegetables, citrus fruit and even rice are the chief crops, but there is also much irrigated fodder grown, such as grass, lucerne and oats. These enable dairy cattle and fat lambs to be reared, a very different type of animal from the rather thin ones which formerly grazed on the natural grassland.

You may think that irrigation is a simple matter, and in a way it is. There are nevertheless still plenty of problems. Engineers must carefully survey the land to build dams and canals. Much capital must be obtained to finance them. Meters to measure the amount of water used and to be paid for must be installed. A new problem is waterlogging and salination. Not only must water be

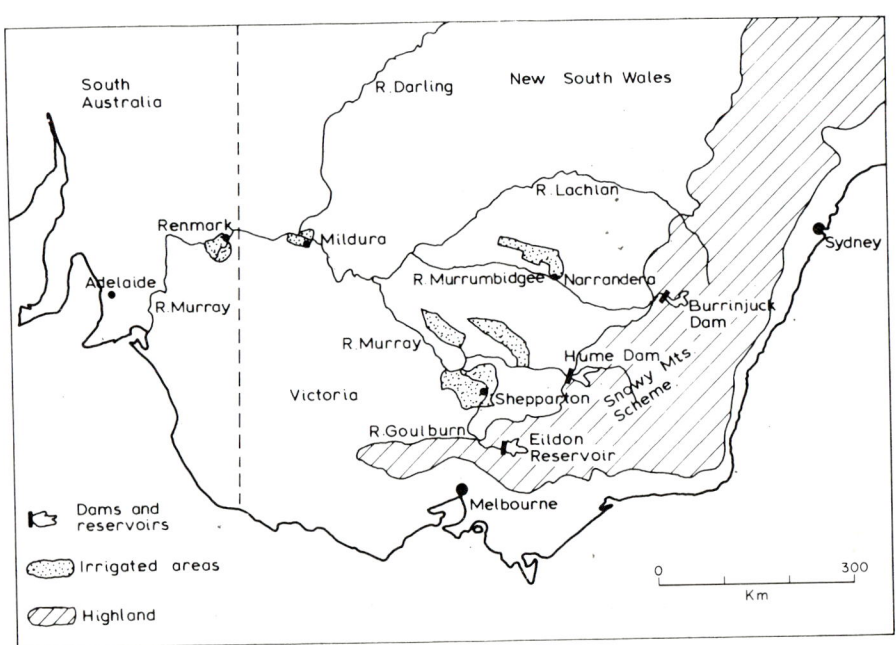

Fig. 2.13. Irrigated areas in the Murray–Darling basin.

52

brought to the land, it must be taken away again as necessary. If too little water is used it has been found that salt from lower layers works upwards. It remains as the surface moisture evaporates, and kills the plants. These problems are being overcome, and in general irrigation is steadily growing and enabling man to make increasing use of the land.

Work to do

1. *On a map of the world mark in the following:*
    i) *The very large rivers.*
    ii) *Latitudes $23\frac{1}{2}°$ N., $23\frac{1}{2}°$ S., $45°$ N., $45°$ S.*
    iii) *Shade in any irrigation schemes mentioned in the text.*
    iv) *Add any others known to you. Give your map a title. Can you add any comments about what this map shows?*

2. i) *Copy the diagram (fig. 2.14) of the man raising water. Using the text, label the diagram.*
    ii) *Write a paragraph explaining how this device works.*

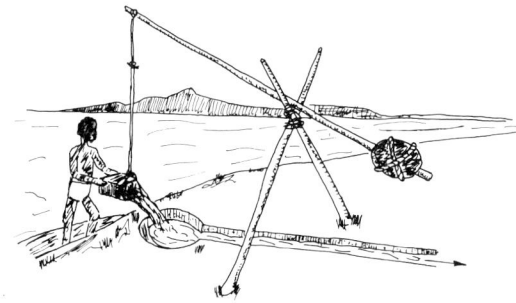

Fig. 2.14.

3. *Write a paragraph entitled 'The problems of irrigation'.*

4. *Draw up a list of the geographical circumstances which give rise to irrigation.*

5. *Copy any one of the maps of irrigated areas; underneath write a paragraph about the crops produced in the area.*

4. Hydro-electric power

Running water and the wind are two great natural sources of power that man has used from ancient times. The water-mill and the windmill are probably the earliest forms of power-driven machinery. A water-wheel is simple and direct (fig. 2.15). Running water can be made to turn a wheel, and this is the basis of the vast modern development of hydro-electric power. Hydro comes from the Greek word for water. Any river can be harnessed to operate a water-wheel, but if there is a natural barrier or waterfall,

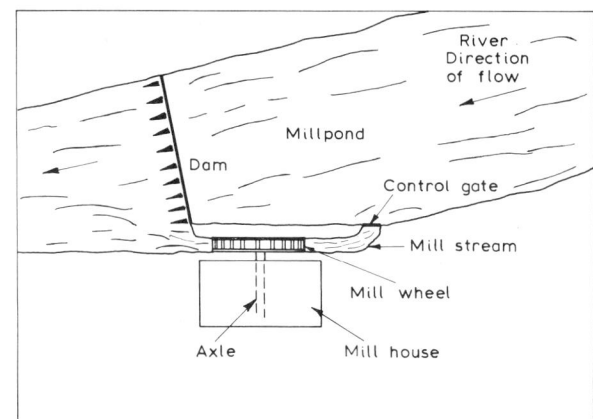

Fig. 2.16. Mill stream and dam.

Fig. 2.15. Mill wheel.

construction is helped. If you examine an old mill on a stream in lowland England you will find it is usually at the side of a barrage or dam across the stream, behind which the water is held up. This was such a common sight in the past that the expression 'as smooth as a mill-pond' has passed into our language. At the side of the pond is a channel, the mill-stream, which leads the water to the mill-wheel. The mill-stream will have a barrier or sluice-gate across it so that the miller can cut off the water to stop the wheel when he wants to (fig. 2.16).

As we shall see later, the modern hydro-electricity station is exactly the same in principle. The water-wheel is called a turbine,

which is only a more elaborate water-wheel, usually housed in a steel casing. The water is led through a tube, and spouts in a powerful jet against the blades of the turbine. The axle, instead of turning the millstones to grind corn, now turns a dynamo, or more correctly, the armature of a dynamo. A dynamo is a piece of apparatus which produces electricity. You can learn exactly how it does this in a science lesson. The basic facts are that when a wire is moved through the influence or field of a magnet a pulse of electricity is generated. The armature is the spinning part which carries many turns of copper wire, and from it the electric current is collected and carried away to wherever it is wanted.

Power in the form of electricity is very convenient, and of course, once the generating station has been built, is very cheap. It is clean to use, producing no smoke as does a coal-fired steam engine, and no dangerous waste material as does a nuclear power station. Above all, it is easy to transport. It can be carried for long distances from where it is made to where it is wanted. For this it is transformed to a high voltage, and is usually carried on overhead power-cables. These cables, on high pylons, sometimes disfigure the countryside, and a modern problem is for planners to find a route for them to which too much opposition is not aroused. In Great Britain all the electricity produced, from water, coal-fired or nuclear-powered stations, is fed into the national grid system, so that supplies can be readily switched from one part of the country to another (fig. 4.6).

Let us study a hydro-electricity scheme in detail. It is not one of the large well-known ones, such as Kariba or the Snowy River, but it is typical of many small schemes in the Alps. It is run by the River Maggia Hydro-Electric Company, a firm which has developed to the full the water-power available from the River Maggia (fig. 2.17). This is one of the smaller streams of the Alps, flowing into Lake Maggiore, and is about 50 km long, dropping some 2000 metres on the way.

High upstream the Sambuco dam has been made across a narrow part of the valley, and behind it is now a large reservoir. From here the water is led through a tunnel in the mountainside to Peccia. Here it drops 400 metres, still in a tunnel, and drives the turbines of the first power-station (fig. 2.18). When the water has spent its force it enters a tributary of the Maggia, but is immediately led again into a tunnel and drops again to a second power-station at Caverno. It again travels through a tunnel, gaining water from tributaries, to the second reservoir at Palagnedra. From here it drops once more to the third power-station on the bank of Lake Maggiore, and the spent water runs out into the lake. In this way nearly all the water of the Maggia and its tributaries is being used to make electricity. This is common in much of the Alps, and often you will see mountain river beds with hardly any water in them, because it is all running in tunnels in the mountainside. From each station power-cables take the electricity over the mountains, to feed power into the Swiss electricity grid, often to drive trains or machines in factories.

We can now consider the geographical circumstances of such schemes, and some of the problems involved. Clearly the more

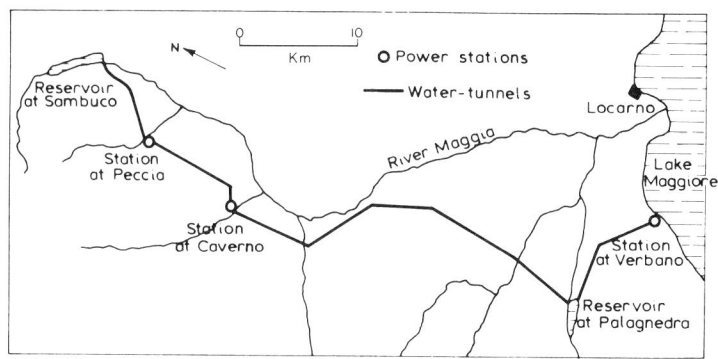

Fig. 2.17. The River Maggia hydro-electricity scheme.

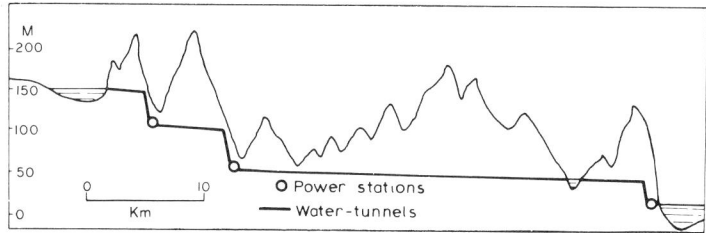

Fig. 2.18. Section to show tunnels: Maggia scheme.

water there is the better, and in general the Alps have plenty of rain. Unfortunately in winter snow falls, and the upper parts of streams are frozen. This reduces the water supply just when it is most needed, and the dams, in which the water does not entirely freeze, must be constructed to ensure water for power all through the year. By spring they are getting empty, and they fill again in summer. Any river which has great changes in its volume, through a dry season, must have some reservoirs constructed, and a river with an even flow all through the year is more convenient.

A dam of the Snowy Mountains Scheme: New South Wales.
Describe this scene in summer. The type of tree is shown on page 179.
(*Australian Information Bureau*)

The narrower the valley, the smaller can be the dam, so narrow or gorge-like sections of the valley are usually chosen. Firm rock on which to base the concrete dams is also needed, and both these conditions are commonly found in mountainous areas. Once the dam has been built, the valley above it fills with water, forming a narrow artificial lake often 20 or 30 or more km long. Sometimes whole villages on the valley floor become flooded, so the inhabitants must be moved beforehand. An example of this is the village of Tignes, in the Savoy Alps in France. In 1954 the old village was covered in water when a dam was built across the River Isère, and a new village was built in a higher part of the valley. Of course the cost of building a new village for the displaced people and providing them with new farmland or other

Power plant at Niagara.
Draw a diagram of this picture, labelling: (*a*) the position of the water-storage reservoir; (*b*) the power plant itself; (*c*) Niagara River. Find these on fig. 2.19. The picture was taken from underneath Queenston Bridge, looking upstream.
(*U.S. Information Service*)

occupations is all part of the cost of building the scheme, and another problem to be solved.

The more the river drops in height, the more often can its water be used, and so wet mountainous regions such as the Alps or the Rockies are major sources of hydro-electric power. Perhaps the ideal situation is where a great flow of water drops suddenly over a natural waterfall, and the Niagara Falls will immediately occur to you. Here the Great Lakes provide a natural reservoir, and the layer of hard rock known as the Niagara Escarpment causes a fall of over 50 metres on the Niagara River (fig. 2.19).

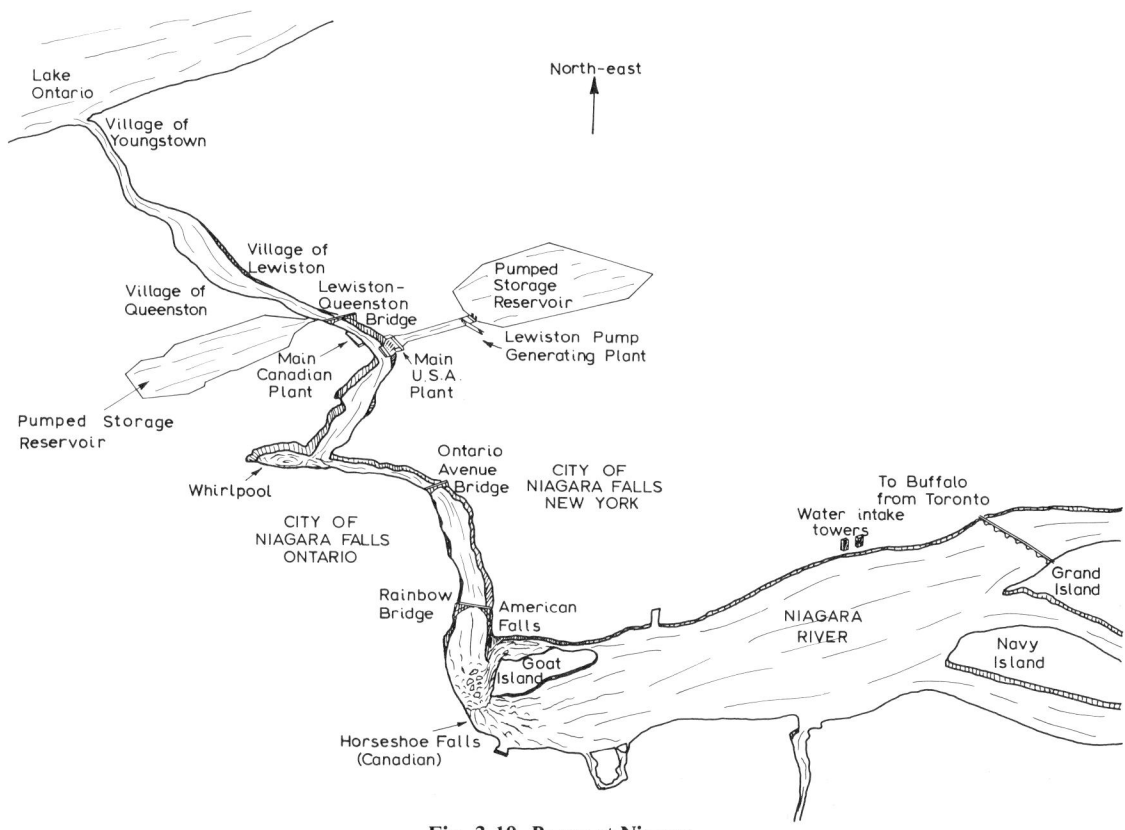

Fig. 2.19. Power at Niagara.

Hydro-electricity production is not confined to mountainous regions. A very large river flowing over more level ground can also be used. The greater volume of water makes up for the smaller drop in height, and a different kind of turbine is needed. One of the oldest and best known of such schemes is at the great barrage near Dnepropetrovsk on the Dnepr, and many other sections of Russian rivers like the Volga are now harnessed, although they flow across quite gently sloping land. The Donzère–Mondragon scheme on the lower Rhône in France is another example of a major scheme on the lower reaches of a river.

We have not yet considered all the problems of the hydro-electricity company. It is no use producing electricity which cannot be bought. If you think for a moment you will realise that there are many rainy mountainous lands, and vast rivers, with no such power stations. The Andes and Himalayas, and the Amazon River are obvious examples. This is partly because the countries concerned are not wealthy enough to find the money needed to build the schemes or buy the product but also because they have no large industries which could use the power produced. The countries of the world with most developed hydro-electricity are thus those which have the necessary rivers, and *also* established industries, or raw material and skilled inhabitants who could work in newly established ones. These countries are at present Canada, the United States, Western Europe, Japan and Soviet Russia. Thus the towns of Northern Italy and Switzerland have many factories using power from the Alps, and in the St. Lawrence lowlands Canada has many prosperous industries using power from Niagara and other stations.

There are some special cases. Large single power-stations are sometimes set up in remote places to exploit particular resources. The Kariba scheme was built because its power could be used in the copper district of Zambia, and the Kitimat scheme in British Columbia was built near the sea, so that its power could be used to smelt aluminium ores. The Volta scheme in Ghana has been built by European technicians and borrowed money to try to start industries in an undeveloped country. Indeed, many under-developed countries would find hydro-electric schemes useful in encouraging industries.

Work to do

1. *In your notebook make a list of the conditions necessary for the successful production of hydro-electricity.*

2. *Write down all the problems which might arise in the planning and building of a hydro-electric scheme.*

3. *Read this section again, and with the help of your atlas complete the following table in your notebook:*

Name of scheme or station	River	Country	Uses to which power is put

4. *Here are some more large, well-known power schemes. Add them to your table.*

Trollhättan—Sweden
Aswan—Egypt
Owen Falls—Uganda
Grand Coulee—Washington (U.S.A.)

Cubatão—Brazil
Snowy River—Australia
Sukkur—West Pakistan
Shasta—California

5. *Write a paragraph describing the advantages of using hydro-electric power.*

Cattle at watering point: Texas. ▶
(*U.S. Information Service*)

Chapter 3

PROBLEMS OF FARMING

1. Farming in Temperate lands

Without food we would starve, and food comes mainly from the land. It is easy for the townsman to forget this when there are always food shops round the corner. Man's first need is the production of food, and farming is therefore of vital importance to him. This is what we shall study in this chapter. Let us look at some facts first. The farm to be described is in many ways typical of farms in this country and in other cool temperate regions.

Little Tosson Farm[1] is in Northumberland, about 6 km west of Rothbury. It is a large, prosperous farm of over 200 ha (fig. 3. 1). The land is in the valley of the River Coquet, and slopes gently down from about 150 m to below 90 m by the river. This gives good natural drainage, without being too steep for machines to cultivate. The soil is fairly light and fertile. There are ample farm buildings (fig. 3.2), which indicate that the farm produces both animals and crops.

There are some 200 Scottish half-bred ewes, producing lambs for sale. There are also about sixty cattle, which are bought mainly in winter, fattened, and sold for beef, and one or two dairy cows for the farmer's own milk. Two-fifths of the land is under permanent pasture and grass for hay, as one would expect on a farm with many animals. Most of the rest grows barley, but there is also one field of oats and turnips. Most of the barley is sold, but some, with the oats and turnips, is fed to the animals. To keep the fields fertile, the farmer adds fertiliser, such as potash, nitrates and phosphates.

In addition, he rotates the crops, that is, changes the crops in a given field from time to time. Different crops take different kinds of nourishment from the ground, and of course animals pasturing when the field is under grass add manure. Fig. 3.3 gives a summary of the main tasks on the farm during the year. As with most

¹ The details of this farm were obtained by the senior girls of Paddington and Maida Vale High School during a field-work course.

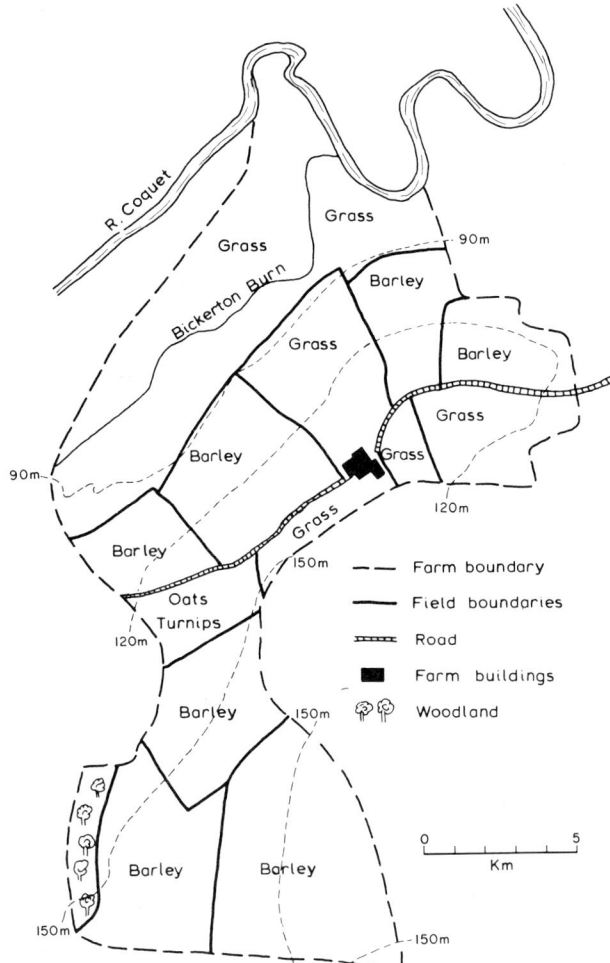

Fig. 3.1. Little Tosson Farm.

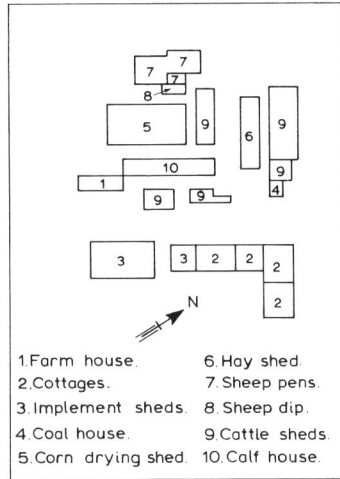

Fig. 3.2. Little Tosson farm buildings.

1. Farm house.
2. Cottages.
3. Implement sheds.
4. Coal house.
5. Corn drying shed.
6. Hay shed.
7. Sheep pens.
8. Sheep dip.
9. Cattle sheds.
10. Calf house.

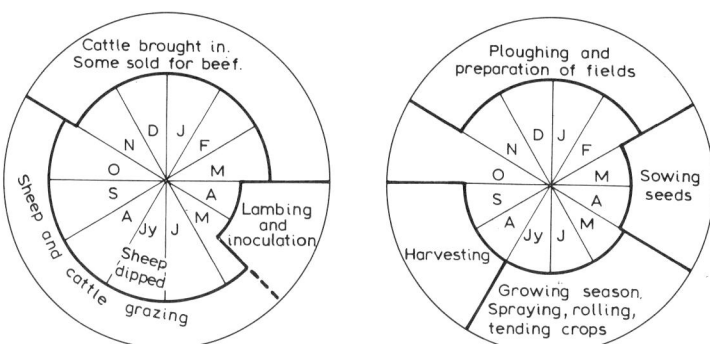

Fig. 3.3. Farm routine: animals and crops.

farms, much machinery is used, particularly tractors and a combine harvester. This enables the farmer to work the land with only two permanent labourers, with three extra at harvest-time.

He buys and sells goods, so markets are important. The fertiliser and seed is bought from West Cumberland Farmers (an agricultural co-operative society), and delivered in lorries. He buys sheep from Scottish and local markets. There is a stock market every fortnight in Rothbury, with special fat-lamb sales in summer. The cattle are bought, and sold when fat, at markets in Morpeth and Newcastle. The wool is sold to the Wool Marketing Board in Newcastle, where there are also millers who buy the barley.

Now let us think of the problems which affect the farmer. You will find that they concern many matters you have studied in geography. You have already read about the weather, for example. The average weather through the year makes up the climate, and clearly this limits the crops the farmer can grow. In these latitudes grain, grass and roots grow best, though there are many

Cattle in stall: England.
Notice the snow on the roof and ground. Refer to fig. 3.3 and find out why and for how long cattle are kept in stalls.
(*Camera Press*)

local variations. You know, too, that although our climate remains broadly the same, the weather can vary from year to year. In a very wet year the barley might not do so well, though a richer growth of grass would partly make up for this. Sheep are hardy animals, but winter storms would be a problem. Cattle are fairly hardy; even so, breeds which can thrive in the cool or even cold weather experienced in the north of England are bought. The farmer is also constantly watching the weather to plan his day-to-day work. A fine dry spell in spring helps the preparation of fields for seeding, and of course in autumn helps the harvesting. You will see from fig. 3.3 how the whole pattern of his year's work is geared to the seasonal changes of the British climate. These details from a weather station at Durham will help you to realise more about climatic conditions in the north-east area.

Mean temperature of warmest month (July)	15° C
Mean temperature of coldest month (January)	3·3° C
Mean daily maximum temperature of warmest month	20° C
Mean daily minimum temperature of coldest month	0·6° C
Extreme maximum temperature	30·6° C
Extreme minimum temperature	−16·1° C
Mean duration of growing season	241 days
Average number of days with frost	67
Average date of first autumn frost	September 30
Average date of last spring frost	May 25
Average duration of frost-free period	127 days
Mean annual rainfall	660 mm
Average number of days with snow	20
Average number with snow lying	16
Average number of days with hail	5
Average number of days with thunder	9
Average number of days with fog at 0900 G.M.T.	26

The farmer must also consider the land itself. The well-drained lighter soils are good ploughland. The heavier, damper land by the river is more often left under grass. He must know about soils and fertilisers. Ignorance of these is one of the biggest problems in less advanced countries. His land must not be too high, since in this country land above 300 metres is too wet and bleak for cultivation, and provides only very rough, poor pasture. Again, this particular farm slopes to the north, and a local problem would be cold winter gales. You will see that the cattle are brought into sheds in winter.

He is also a practical biologist. He must know a little about the different varieties of barley, oats and grass, and which ones are suited to his land. He must also know animal husbandry, that is, how to look after his cattle and sheep. Both plants and animals suffer from diseases, and he must know how to diagnose and prevent them. In advanced countries there is an agricultural advisory service to help farmers. One of the most important branches of the United Nations Organisation is the F.A.O., or Food and Agriculture Organisation, which tries to give such technical help to countries in need of it.

Spreading fertiliser in spring: England.
(*Camera Press*)

Sheep in orchard: Kent.
(*Camera Press*)

Farming is now increasingly mechanised, and you probably know the main implements pulled by the tractor. There are other machines in the farm buildings, to dry corn, chop and mix food and so on. All these cost money, so the farmer must have capital, and be a mechanic to repair them.

Above all, the farmer is a business man. He must organise the work efficiently, and have things ready at the right time. The seed-drill must be mended, if necessary, before seed-time, and it is no use deciding a fine day is suitable for spreading fertiliser if he has run out of supplies. He must think of jobs for the men to do indoors in bad weather, and arrange in good time for his harvesters. In addition, he must buy and sell wisely. He must judge the

right price, particularly at stock sales; the wool and barley are sold at prices agreed in advance.

We must now consider how far this kind of farming is typical of all farms. We have described a commercial mixed farm, that is it grows crops and rears stock for sale, and this is common in many temperate lands. Of course, there are variations from this. You will have thought of the dairy farm, which concentrates on producing milk, butter and cheese, and has sometimes much pasture land. Such farms also often keep pigs and poultry, though this is another specialised occupation nowadays. There is the market-garden, which grows vegetables or flowers. Such farms are usually quite small, where intensive care can be given to the plants, which are of high value. These and dairy farms tend to be found in the neighbourhood of large cities, so that the perishable produce can be put quickly on a large market. This is not always the case, however. Another specialised farm is the fruit farm, though this is not necessarily one large orchard. The Kent farmer pastures sheep between his apple or cherry trees, and the Worcestershire fruit farmer may also have a few cattle or grow vegetables.

The largest and most important variations on the mixed farm are those which mainly grow grain, and those which mainly rear animals. Both are found particularly in the temperate grassland regions, where population is less dense, and the use of the land is less intensive. You could turn now to page 200 to read about a wheat farm in Canada, and you have probably heard of the great cattle ranches of the United States and the Argentine, or the large sheep farms of Australia. Even in Australia, however, there are plenty of farms which rear sheep and grow wheat (fig. 3.4). In all these different kinds of farms the problems are much the same, of land, climate, care of plants and animals, and of marketing.

The crop and animal farming we have described is found in temperate latitudes. The location of the temperate grasslands is shown on fig. 8.15. The other main regions where this kind of farming occurs are the cool, damp lands of these latitudes, nearer the coasts. The chief areas are north-west and central Europe, north-east United States and south-east Canada. Smaller areas are on the west of the United States, in the states of Washington and Oregon; in British Columbia, in Tasmania and New Zealand. There are many other temperate crops which we have not mentioned, and not all the farmers concentrate on selling their produce, though most of them do. In parts of Europe particularly farmers produce a variety of food for their own consumption, and are thus less dependent upon their cash income, but we will consider this kind of farming in the next section.

Fig. 3.4. A wheat–sheep farm in New South Wales.

Work to do

1. *Write a description of the typical weather of the four seasons in Britain. If you keep school weather records you can use these to help you.*

2. *Study the weather information about Durham (page 62) and answer the questions.*
    i) *In our latitudes plants stop growing when temperatures are at 5·6° C or below. Temperatures higher than these form the growing season. For about how many months does the growing season last?*
    ii) *Name the months in which frost is likely to occur.*
    iii) *For how long will farmers' work be hindered by snow lying on the ground?*
    iv) *Days with thunder often mean heavy rain storms. In what season will these be particularly harmful to the farmer's crops?*
    v) *Name the highest and lowest temperatures experienced in this area.*

3. *Write a list of the main implements a British farmer uses.*

4. *What steps can the farmer take to prevent diseases: (a) in stock; (b) in crops?*

5. *The following table shows the main imports of food into the United Kingdom, in millions of pounds value:*

Meat		Fruit and vegetables		Dairy produce and eggs		Wheat	
Denmark	125	South Africa	47	N. Zealand	79	Canada	69
N. Zealand	100	Australia	27	Denmark	39	Australia	49
Eire	64	Italy	26	Eire	28	Netherlands	22
Australia	40	Netherlands	23	Netherlands	18	France	21
		France	20				
		Jamaica	12				

    i) *Draw five separate graphs to show the import value of these products.*
    ii) *Make a list to show which fruits and vegetables you know to come from the various countries mentioned.*
    iii) *Make a list of other foodstuffs or beverages you know to be imported, naming the countries from which they come·*

2. Farming in Tropical lands

The southern part of Europe is not tropical. It is, strictly speaking, a warm temperate or sub-tropical region, but it is something of a halfway house between the farming we have just studied and the farming of lands properly within the tropics, so it is worth pausing on our journey south, so to speak, to consider its agriculture at this point. These lands take their name from the Mediterranean Sea (fig. 8.9), and there is more detail of their climate and life on page 192.

There are small farms worked by the owner and his family, who naturally take a very great interest in them. This is called peasant farming. Much more of the produce is for the farmer's own use, though in these days he is far from being self-sufficient. When the whole of the family's needs comes from the farm we speak of subsistence farming, of which you will hear later. As the holdings are small, they are very intensively cultivated. Two and even three crops can be found on the same land. Thus there could be rows of vines, with lettuce or beans between them, and lemon or peach trees at intervals. There are also many large estates, as in southern Italy, where sometimes the labourers have a plot of land on which they grow their own food. In remote parts, such as the Greek islands, there are still peasant farmers who grow their own wheat, vegetables and fruit, make their own olive oil, rear their own hens and goats, and even spin wool from their sheep, but in general nowadays more and more is sold to the townsfolk, and the increasing number of visitors.

Many lands round the Mediterranean are hilly or mountainous, usually of limestone rock, which holds little or no surface water, and is often bare of soil. By hard work over the ages many of the hillsides have been shaped into terraces, which can be cultivated and which can hold soil. Water for irrigation on these hillsides is difficult to obtain. The sheep and goats which can live on these hills have overgrazed the land (page 9), and what might be quite

Gathering olives: Greece.
Read also the description on page 194.
(*J. Allan Cash*)

good forest land is as a result very rough scrubby ground of little value. Where water can be obtained for irrigation the sunny Mediterranean lands can be very productive, but there are still parts of Greece, southern Italy and north Africa where the people are very poor. There are in fact two kinds of agriculture. There is the old style, such as is mentioned in the Bible, where man grows wheat, olives and vines, and keeps sheep and goats, and the new style, where heavier natural rainfall or irrigation systems help man to produce oranges, lemons, peaches, pears, walnuts, grapes, lettuces, onions, aubergines, water-melons, tomatoes, maize and a host of other crops.

This varied peasant farming is found only in the lands round the Mediterranean itself. In Central Chile, on large estates, the workers have their own plots of land as in Italy. In the other main areas, California, South Africa and Australia, the same crops are grown, mainly fruit, vines and vegetables and sometimes wheat, but they are usually on specialised farms on a highly commercial basis. You will be aware of the enormous fruit production of the Central Valley of California, and a local problem here is to make provision for the hordes of migratory workers who make their way from south to north, picking the fruit as it ripens.

In the tropics proper, and the well-watered sub-tropical lands near them, there are broadly three kinds of agriculture, and it is difficult to realise the importance of the first kind, the subsistence rice farming of South-East Asia. Perhaps 1000 million people, between a quarter and a third of the whole population of the world, depend, many of them quite directly, upon rice. This yields more nutritious food per acre than any other grain crop, and has a strong husk which makes it store well. As you will see on page 185, the conditions are well suited to rice growing, but probably the safest explanation of its existence is that over many centuries this is the kind of plant that the people have developed. Lowland rice needs much water, and the plant actually grows in flooded fields. The areas with the highest proportion of land under rice are

therefore the level deltas and flood-plains of the great rivers of South-East Asia (fig. 3.5). At certain seasons, when the crop is well up, you can see on a journey little else but bright green padi, as the rice is often called, making a monotonous level landscape interrupted only by farms and villages. In the lower Ganges valley, for example, such scenery continues for hundreds of km; there is village after village, dependent for its food on the rice grown near by. All these people depend on the land: if the crop

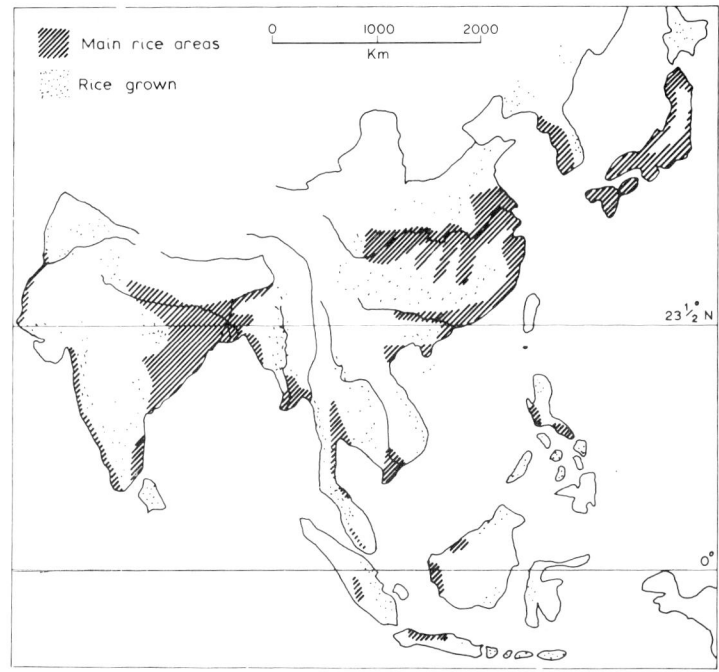

Fig. 3.5. Rice growing in South-East Asia.

Terraces for irrigated rice: Philippines.
(*Aerofilms*)

Ploughing flooded padi fields: Japan.
What are the differences between this and an English ploughing scene?
(*J. Allan Cash*)

fails, they starve. In some mountainous lands by enormous work the people have made terraces for irrigated rice, producing the remarkable landscape shown in the picture.

The basic processes of rice cultivation are the same as for any other grain, of ploughing, raking and levelling the fields, harvesting and threshing. There is much extra work in addition. The walls round each field, to enclose the water, must be kept up, and the drainage channels tended. There is an extra process of transplanting the rice from small seed-beds, which increases the yield. All this is done with the simplest of implements. The soil is dug with a plough, drawn by an ox or water buffalo, and smoothed

with a roller or harrow. The rest of the work is done by hand, using a hoe for weeding, a sickle for reaping and a flail for threshing. In Japan small motor cultivators, suited to the small fields, are now increasingly being used.

Each farmer's holding is very small: a hectare or less is common. Even in China, where the holdings have been collectivised, there are the same number of people to work communally on the same amount of land. This means it is used very intensively. By giving a great deal of care to the small patches of land, enough food can be obtained. It is very useful that in these warm regions, with careful cultivation, two and even three crops a year can be got from the same plot. Rice needs a temperature of 21°C to grow, but four months is enough from planting to reaping, so where the temperature remains above this all the year, the land can be used three times over. In cooler parts other crops are grown instead of rice in the winter. Fig. 3.6 shows the crops grown on two small fields in Japan over two years.

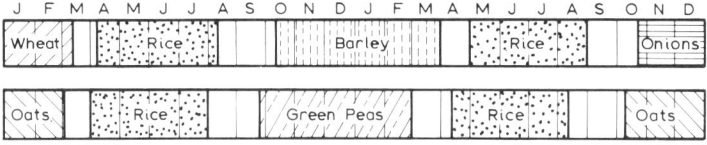

Fig. 3.6. Crops on a Japanese farm.

Rice, as fig. 3.6 shows, is not the only crop, but it is by far the most important. Vegetables, other grains, some fodder crops for animals—though these are few—and spices for flavouring are some. Not all South-East Asia can grow rice. In northern China the climate is too cool, and in much of India too dry, and here wheat and millet are the chief cereals grown. In all these places the same care must be given to the land, to get as much food as possible.

The problems of agriculture are still the same, of land, water, crops, fertiliser and knowledge. In a way, the problems here have been solved. This system of rice growing has evolved over centuries, and in these crowded lands it works. The problem is that it cannot be developed much further. There is no more land available, and if population increases the amount of food per head is less. This is partly remedied by emigration (page 164), but for millions of Indians, Chinese and Japanese who cannot or do not wish to emigrate there is not much prospect of an early solution.

We have discussed subsistence rice farming, the first of the three broad types of tropical agriculture. The second kind of agriculture is the most primitive known, and is called shifting cultivation. It is found mainly in the equatorial forests (page

174). It occurs in the Amazon Basin and central America, in parts of the Congo Basin, and in the islands of the East Indian Archipelago. It supports very few people, and is steadily declining.

It is practised usually by a tribe or a large family group. An area of forest is selected, not too hilly and not too flat, and the land is cleared. This is the hardest work, as the only implement is the axe. Sometimes large trees are heavily lopped or ringed so that they remain bare stumps. The branches, leaves and any other refuse are then burnt; naturally a period just after the rainiest season is selected. The ashes put some nourishment into the soil, which at first, being made of leaf mould, is fairly rich. There is no ploughing; these people have only a hoe or a piece of wood, called a digging stick. Holes are dug in the ground, and the seeds inserted. This in some cases is all the attention the plants get; in others some weeding is done, and the ground may be hoed up into mounds for certain crops like yams, which are rather like large potatoes. Another common crop is cassava (page 173), but many others, such as maize, bananas, peppers, beans and ground-nuts, are grown in different countries. Fig. 3.7 illustrates this.

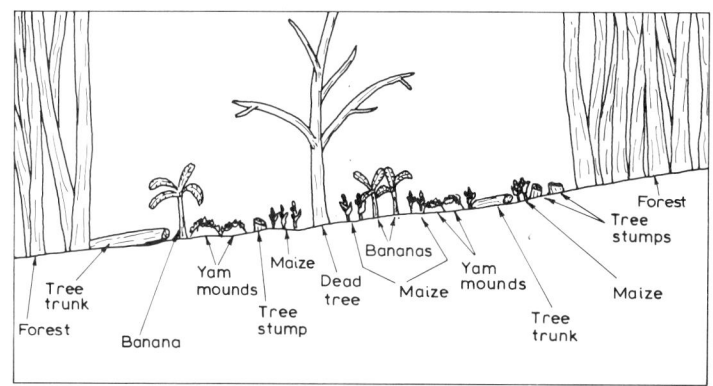

Fig. 3.7. Crops in a forest clearing.

There is good growth at first in these hot, wet conditions, but the soil soon becomes impoverished as the heavy rain washes out the plant foods. The land is abandoned after two or three years, and a fresh piece cleared. The old land rapidly becomes overgrown, and will not be used again for twenty or thirty years. It never really returns to the original forest described on page 172. You can see that if there is sufficient forest available a tribe may stay in one place and gradually clear bits of forest in the area. The true shifting cultivators do not bother to do this. Their houses are very simple, and when they need to find new land they just move on and build new huts somewhere else.

The main problem is the heavy leaching of the soil in the tropical conditions. Any vegetable matter rots down easily, but as it is then soluble, the rain soon washes or leaches it away. As the tree cover has been removed, there is little protection from the rain (page 9). The burning is also wasteful. Much plant nourishment literally goes up in smoke, and the ashes produce less than if the green leaves were slowly rotted down and then added to the land. There are no animals, either, to add manure. You can see that this is a very simple form of agriculture. It is practised only by the most primitive peoples, and it requires a great deal of land to support a small population.

Most native peoples in the tropics have a more advanced and productive system of agriculture, which is the third of the three kinds we mentioned. It is a subsistence economy, that is, farms or villages are almost entirely self-supporting, though this is changing, as you will see. It also shows traces of shifting cultivation. It is settled farming, growing a great variety of tropical crops, on fairly small patches, often keeping animals, and using the principles of rotation and fertilisation.

The quite densely populated lands of West Africa are farmed in this way. Much of the forest is less dense than the true equatorial forest, and there is a marked dry season. Many of the villages are in clearings, and each family has a small area of land round its house, called its compound land. This is carefully dug with a hoe, and fertilised, using kitchen waste, rotted leaves and some ashes. It grows mainly yams, with many other vegetables and fruits, such as we have mentioned, often planted between them. The family also has a hectare or so of land in the forest, which has been cleared by previous generations. This grows yams or cassava for two years, and it is then rested, or fallowed, for three years. It is rapidly covered by 'bush' or jungle, which is cleared and burnt for the next planting. There is still the problem of leaching, and the crops from the forest patches must be carried home, sometimes several kilometres.

There are also many tree crops, which have become increasingly important. The oil palm grows wild in West Africa, and of course the trees are left standing when land is cleared. Many farmers also plant more (page 174). In West Africa, too, the cacao tree has been so successful that many small farmers have planted whole groves of them. There are also fruit trees you know, such as oranges and pears, and others less familiar, such as mangoes, avocados and bread fruit. These subsistence farmers therefore now have cash crops, which form an increasing part of their income. Cocoa and palm oil is sold overseas, and maize, yams and vegetables are taken to market in the many towns which now exist. A problem for the farmers is that their income from cocoa and palm oil depends on world prices, and the cacao trees have suffered severely from a disease called 'swollen shoot'.

This kind of native agriculture, cultivating fairly small plots of land round a village, usually by hoe not plough, is common also all over the grasslands of Africa (page 178), but the crops are different, and there are more cattle (fig. 3.8). As the rain is less, millet and maize are the chief grains, but sugar cane, sweet potatoes, beans, peas and other vegetables are grown. Just as in the forest lands, there are now some crops for sale; many African farmers produce cotton or ground-nuts as a cash crop. Coffee trees grow well in these lands, and a few are grown by natives on their plots, but most are grown on plantations, as their cultivation requires special knowledge.

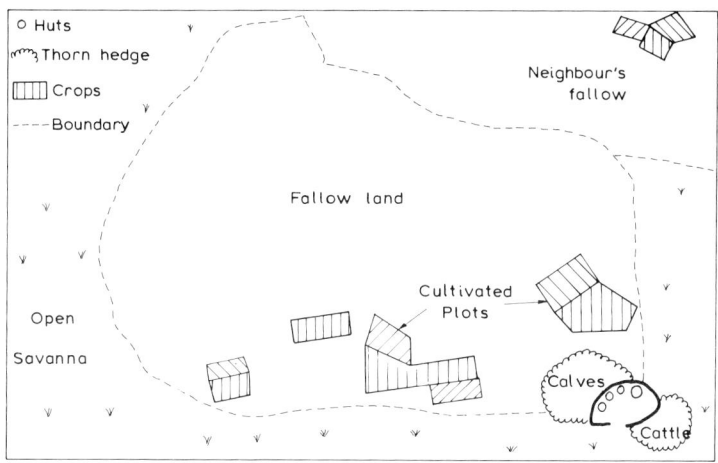

Fig. 3.8. Cultivation in Tanzania.

Fulani cattle at water hole: Northern Nigeria.
Describe the landscape. What is the vegetation? How do the cattle differ from ours? Why are they thin? What will happen to the water hole eventually?
(*J. Allan Cash*)

In addition to those you already know, there are some further problems in this farming. The savannas have a long dry season, so only one crop a year can be obtained. During the dry weather the land becomes baked hard, and cannot be dug until the rains come, although this is just when all the crops must be planted. As in some monsoon lands, the rains can be very irregular. The cattle must be kept out of cropland, though after the harvest they can pasture on the stalks, and their manure is very important. There must be enough cattle to eat the grass and to keep it short, but not too many to cause overgrazing. In the dry season they may have to be moved many miles to find pasture, usually on the damper land by rivers. There can be difficulties, too, between cattle-owning peoples, who are most interested in pasture, and others who want more land to grow crops.

The small-scale native farming we have described is also wide-spread in Central and South America, but there are other types. There is the intensive irrigated farming of oasis lands, and the nomadic herding of the deserts themselves. In the Himalayas and Andes there is animal rearing somewhat similar to that of the Alps. There are also large areas of western commercial farming, such as the white-run mixed farms of the highlands of Kenya, and the cattle-ranches of Queensland. The most highly specialised commercial farms in the tropics are plantations, described in the next section.

71

Work to do

1.

Crops	Winter crops	Summer crops
Grain Vegetables Fruit		

Copy the table, and complete it for farming in the Mediterranean lands.

2. Write a description of what man has done to the landscape shown in the picture of terraced ricefields.

3. Write an account of rice growing in any one delta area of South-East Asia.

4. Using the index of the book to help you find information, write a paragraph each on:

i) the oil palm;
ii) small-scale native farming in West Africa;
iii) soil problems in tropical lands.

5. 'The big huts, with their conical thatched roofs, were grouped neatly round a small square which was shaded with groups of young eucalyptus trees. In this square was the market; in the patchwork of light and shadow under the slim trees the traders had spread their wares on the ground, and around them thronged the villagers in a gesticulating, chattering wedge. The wares offered for sale were astonishing in their variety. There were freshwater cat-fish, dried by wood smoke and spitted on short sticks. There were great bales of cloth, some of it the highly coloured prints so beloved of the African, imported from England: more tasteful was the locally woven cloth, thick and soft. Among these patches of highly coloured cloth were an odd assortment of eggs, chickens in bamboo baskets, green peppers, cabbages, potatoes, sugar-cane, mangoes, pawpaws, lemons, yams, lovely raffia-work bags, leopard skins, calabashes full of palm wine and old kerosene tins full of palm and groundnut oil.'

i) Of what part of Africa was this description written? What words give clues?
ii) Make a list of the fruits mentioned. Underline those which are grown only in tropical lands.
iii) How are palm oil and ground-nut oil obtained?
iv) Which local industries are mentioned?
v) Draw a sketch of the scene.
vi) How does this market scene differ from a market scene in England? In what ways is it similar?

3. Plantations

Plantations are areas of commercial crops growing generally in tropical lands. As the crops are grown for sale, they are usually produced on a large scale, and their care involves a great many workers. The need for labourers in plantations has given rise to large population movements. It is thought that in the days of slave labour some 5 million Africans were transported, mainly to the sugar plantations of the West Indies. Fortunately, most workers have gone to plantation areas voluntarily. Indians have sailed to Sri Lanka (Ceylon) and Malaysia; Italians and Negroes to Brazil, and Chinese to Indonesia and Malaysia. All these people have left their own countries when work and food were scarce to find work in plantations, either directly as plantation labourers or setting up shops, helping in trade, building roads and railways, or working at other occupations (page 164).

In the early days of plantations the areas of land where they were started had been largely neglected for cultivation and had few inhabitants. Most plantations were started by companies, who not only planted the crops but were responsible for building roads and railways, providing clean water, setting up houses, schools and health services for their workers. Even now the social welfare of the people of Assam and Sri Lanka depends on the success of the tea plantations, that of the Cubans depends on sugar and tobacco plantations, while the cultivation of rubber enables nearly half of Malaysia's population to have the highest standard of living in south Asia. The supply of products from plantations bulks large in world trade and in meeting the needs of countries for raw materials and food. Thus plantation agriculture is very important.

Let us look at a rubber plantation. Rubber is made from the juice or latex which is found in the tissues underneath the bark of certain evergreen trees generally known as rubber trees. You probably already associate the name of Dunlop with rubber, from

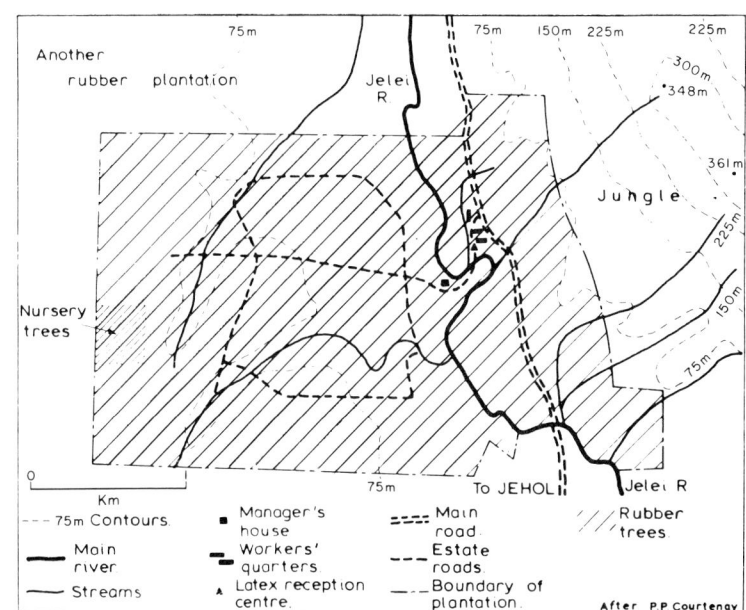

Fig. 3.9. Kuala Jelei rubber plantation.

tyres to tennis and golf balls, so it will not surprise you to learn that the Kuala Jelei rubber plantation is one of a number owned by Dunlop Malayan Estates Limited. Fig. 3.9 shows you detail of the plantation. The plantation is 681 hectares in area. It is crossed by the Jelei River, which with its many tributaries drains the region and provides the estate with a plentiful water supply. You will notice that the rubber trees are grown on the lower land. Although rubber has been grown on the plantation since 1927, most of the old trees have been replaced by newer trees planted between 1948 and 1959. 450 trees are planted per hectare, but when they reach full size the number is thinned to 325 trees. These yield nearly 2000 kg of rubber per hectare yearly.

Plantation factory: pails of latex: Sri Lanka.
(*J. Allan Cash*)

Apart from an English manager, the workers at Kuala Jelei consist of Asians. Eighty-nine men and thirty-two women are Malaysians, forty-two men and sixty-three women are Chinese, and seven men and five women are Indian. The worker who collects the latex from the tree is called a tapper. By means of a large knife he removes a very thin sloping strip of bark from one side of the trunk, so that the latex wells up. It then runs down along the cut, through a small metal spout which the tapper sticks into the tree and into a cup fastened below the spout. Each tapper deals with about 400 trees daily. He starts cutting at 6 a.m. daily and

continues until 9.30 a.m. By the time he has finished cutting his last tree the cup on the first tree cut will be filled with latex. This he starts collecting, pouring the latex from the cup into a large can. He tries to finish emptying all the cups before the heavy rains of the early afternoon begin. When he has carried his heavy cans to the collecting lorry he has finished his day's work, usually by 2 p.m. In addition to money wages, the plantation worker is provided with free housing, water, medical services, sports-field, a temple and a mosque, schools and garden land for vegetable growing.

Ideally 3 per cent of the area under rubber should be replanted yearly. To do this replanting, some of the workers look after the young seedlings in the nursery area, and others cut down old trees and replace them by the nursery seedlings. Other workers keep the plantation free from weeds, or drive the collecting lorries. Tapping is perhaps the most skilled occupation, for the cut in the tree bark must not be more than one millimetre deep; if the cuts are deeper the bark will not grow again.

About 20 per cent of the latex collected on the Kuala Jelei plantation is processed into crêpe rubber in a mill near the reception centre. The rest is processed into concentrated liquid latex or liquid rubber. This is transported by road and rail tankers to Singapore for export to the United Kingdom, New Zealand, Australia, Japan and Europe. In the rubber factories of these countries it is made into a large number of goods, including the tyres and balls we have mentioned.

There are many plantations like that of Kuala Jelei on the better-drained lowlands of western Malaysia. Here the consistently high temperature of 21°–27 C with rainfall of about 2000 mm falling evenly throughout the year enable the rubber trees to flourish. Such a climate encourages other trees to grow also, and many of the areas now in use as rubber plantations had to be cleared of forest first. The world demand for rubber has encouraged man to establish rubber plantations elsewhere (fig. 3.10), as in West Africa and Brazil, but the plantations of South-East Asia are by

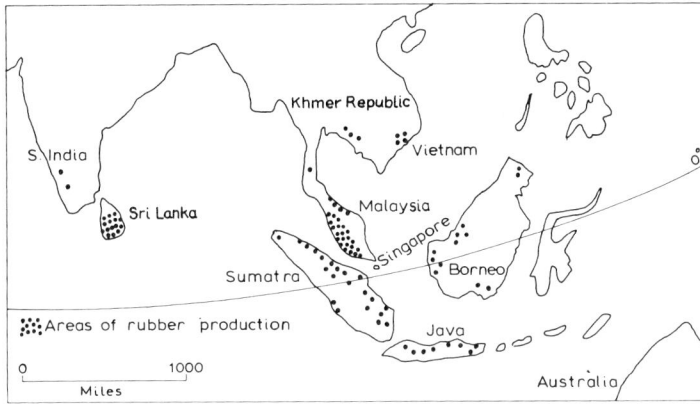

Fig. 3.10. Rubber areas in South-East Asia.

far the biggest producers of this vital tree crop, a raw material for industry.

Plantation crops include not only raw materials like rubber but food crops like sugar, bananas, palm oil (page 174), coffee and tea. About 90 per cent of the world's tea is produced on plantations (fig. 3.11), although commercial tea growing occupies a much smaller area than that taken by rubber plantations. We in our islands are the world's heaviest tea drinkers; the New Zealanders and Australians come second and third. Most of our tea is imported from the sub-continent of India and Sri Lanka, but tea of very delicate flavour and scent comes from China.

Loading bananas on a plantation: West Indies.
(*J. Allan Cash*)

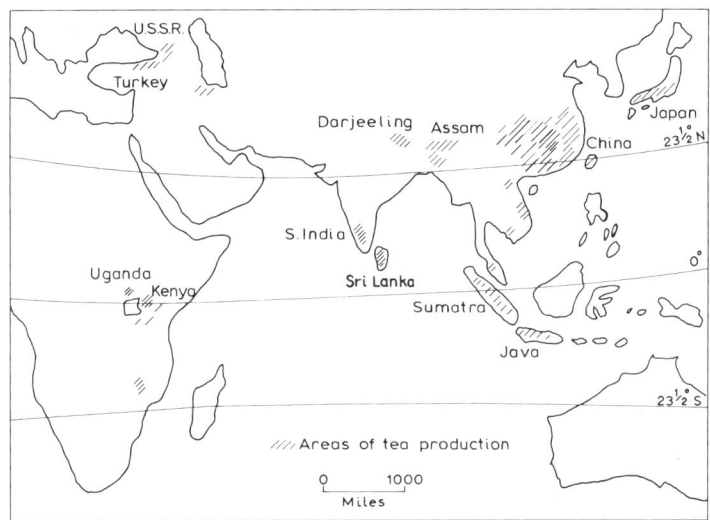

Fig. 3.11. Areas of tea production.

Tea is an evergreen plant of sub-tropical areas. If a tea plant in India is left to grow naturally it becomes a 15-metre tree with branches bearing large, shiny leaves. In plantations the plant is pruned constantly so that it grows only to the size of a bush about 1 metre high. This constant pruning not only encourages the leaves to grow abundantly, so that they can be picked or plucked over long periods, but results in small leaves and shoots which give the best-flavoured tea.

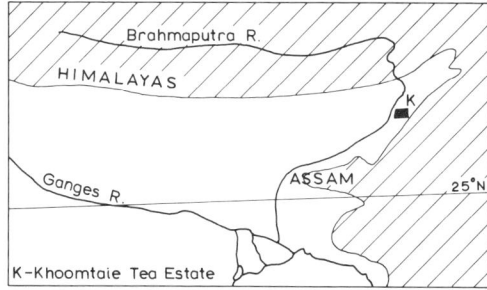

Fig. 3.12. Location of the tea plantation.

The Khoomtaie Tea Estate or plantation belongs to the Assam Company Limited, who own nearly 4000 hectares of tea-growing land in Assam (fig. 3.12). The plantation covers 1500 ha, but some of this land still needs to be cleared for tea growing (fig. 3.13). Although tea requires a well-distributed annual rainfall of well over 1250 mm, it will not live if its roots lie in waterlogged soil. For this reason it is often grown on hillsides where the water

◀ **Tea picking on plantation: Assam.**
Why do the pickers carry their baskets on their backs? How do they support them? What height are the tea bushes? Why are they kept small? What grows on the hillside besides tea?
(*J. Allan Cash*)

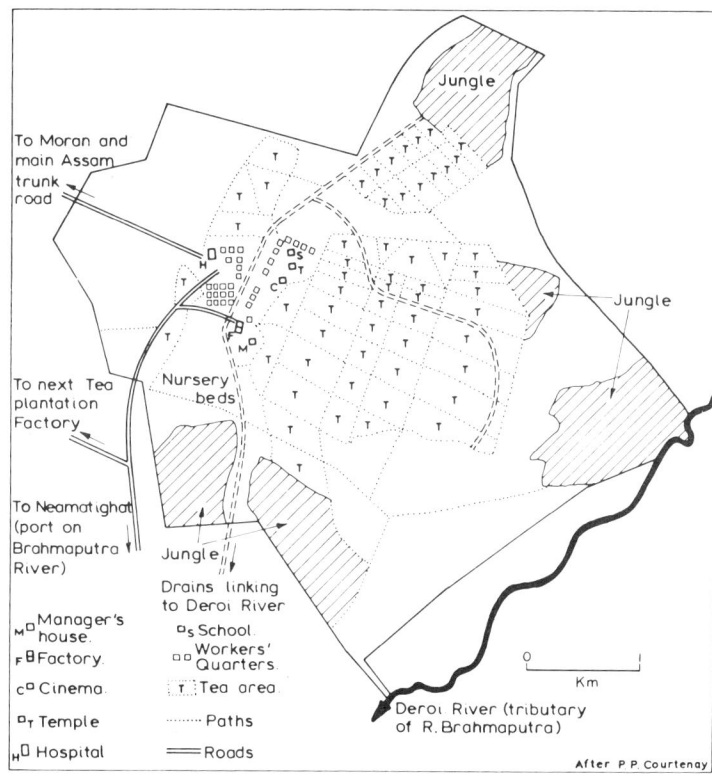

To Moran and main Assam trunk road

To next Tea plantation Factory

To Neamatighat (port on Brahmaputra River)

Nursery beds

Jungle

Drains linking to Deroi River

Jungle

Jungle

Jungle

M◻ Manager's house.
F◻ Factory.
C◻ Cinema.
◻T Temple
H◻ Hospital

◻s School.
◻◻ Workers' Quarters.
T Tea area.
····· Paths
═══ Roads

0   Km   1

Deroi River (tributary of R. Brahmaputra)

After P. P. Courtenay

Fig. 3.13. The Khoomtaie tea plantation.

drains away naturally. The Khoomtaie plantation lies on the flat lowlands of the Deroi River at about 100 m above sea-level, so it has to be drained artificially by means of ditches through which the excess rain flows away to the river. The temperatures in this area average over 27° C throughout the year, so there are no frosts and the tea flourishes.

The manager is English, but the 783 men and 727 women working on the plantation are Indian. Over half of these are pluckers, who work eight hours daily from the beginning of April until the beginning of December picking off the smallest leaves and putting them into baskets carried on their backs. When the leaves have been collected they have to be dried and rolled in the factory on the estate, before being packed into chests sent by road to a port on the River Brahmaputra. From here the tea is sent by boat to Calcutta, and thence by sea to the United Kingdom.

The other workers of the plantation are always busy weeding, manuring, clearing out ditches and clearing light jungle so that the tea area can be extended. From December to April, when the weather is a little cooler, plucking stops, so the pluckers are set to help with these and other tasks, such as house and road repairing, planting seedlings in the nursery and replacing worn-out bushes with young plants ready to be taken from the nursery. The factory is closed at this time so that the machinery can be overhauled. At this time of year the workers have less to do, so they work shorter hours. Their money wage may then be less, but they still have free welfare services, such as housing, schools, medical services, clubs, canteens, sport facilities, cinema shows and centres of worship, like those of the rubber-plantation workers.

You can see that the problems of plantations include the initial outlay of a great deal of money. Since the plantations may be located away from large towns, the owner feels responsible for making the lives of his workers as pleasant as possible, so he provides houses, amusements, doctors and nurses for small hospitals, and places for worship. Indeed, many plantation owners really build large villages or small towns on their estates. The cost of building and maintaining these is often so great that groups of people form companies to establish large plantations, since individuals could not provide sufficient money. Not all plantations prosper. Sometimes insects or diseases attack the plants so that large areas of plantation are ruined. During the Second World War many plantations in Malaysia and Indonesia were abandoned. They were soon over-run with weeds and returned to natural jungle, and some have not been re-established. Since plantations crops grow only one crop, the price received for it is important; if too much of any one crop is produced prices fall and plantation owners can be ruined. This has happened in the case of the coffee plantations of Brazil and Kenya. The transport of plantation crops contributes a large share in world trade. When this is flourishing the plantation owners gain a satisfactory return or profit on all the money they have invested in their plantations.

Work to do

1. *Study the map (fig. 3.9), then answer these questions:*
    i) *What is the name of the main river?*
    ii) *How many tributaries has it?*
    iii) *Where do most of these tributaries rise?*
    iv) *What is the height of the highest hill?*
    v) *At what height does the rubber plantation lie?*
    vi) *How far has a labourer to walk from his house to the rubber nursery?*
    vii) *What is meant by a latex reception centre?*
    viii) *To what towns does the main road lead?*
    ix) *What grows all around the plantation?*

2. *The figures below give the percentage of world production of rubber for the most important producing countries:*

Indonesia	40	Sri Lanka	5	Liberia	2
Malaysia	36	Vietnam	3	Others	8
Thailand	6	Brazil	2		

    i) *Draw column graphs to show the world production of rubber, using a scale of 2 cm for 10 per cent.*
    ii) *Which of the countries named are not in Asia?*

3. i) *Describe the climate required for the cultivation of rubber.*
    ii) *State two of the disadvantages of this type of climate for the workers.*
    iii) *Describe the daily work of a rubber tapper.*
    iv) *Name a port from which rubber is exported.*

4. *Describe the Khoomtaie Tea Estate, using the map (fig. 3.13) to help you.*

5. *Write a short essay entitled 'The problems of plantations'.*

Coal fired power station: Warwickshire. ▶
(*Camera Press*)

Chapter 4

PROBLEMS OF INDUSTRY

1. Industry and the landscape

When you are working hard you are said to be industrious. Strictly speaking, that is all industry means—work. Nowadays it has come to mean work which is directed to making things. Even Stone Age man chipped at flints to make axes or arrowheads, and there was most flint-chipping where most flints could be found, a very early example of industrial location. As civilisation developed, people began to specialise in making one sort of thing, so that they became better at it than others. The craftsmen of the Middle Ages were such specialists, who were usually grouped together in certain streets of towns, where could be found cloth-makers, armourers, tanners and so on. They might have simple machines, such as looms, or bellows to blow their fire, but nothing very elaborate. The work was usually done by one man, with perhaps an apprentice, working in his own shop with his own labour.

Another important development in industrial history was the gradual use of power provided by nature. Man learnt to use wind or water power, and this enabled him to make more goods. The biggest step forward was during the period called the Industrial Revolution, which developed in this country about two and a half centuries ago. Not only were new machines invented but they were larger ones which had to be driven by power. At first this was from falling water, but later from the steam engine, fired by coal. These large machines were put in one building, a factory, so that all could be driven by the same engine. Today manufacturing industry is one of the most important ways by which man earns his living. Factories now consist of many different buildings, though some industry is still carried on in fairly small workshops. There are also some very large works, such as those which make steel or chemicals, which could hardly be called buildings. All these activities make up manufacturing industry, and we shall consider some of its problems in this chapter.

Sometimes the word industry is used rather freely for almost any of man's activities. You may see mentioned in the papers the farming industry or the fishing industry. The mining of minerals from the ground is often called extractive industry. People even write sometimes of the transport industry, or the hotel industry, but this is not quite correct. Activities such as these are properly called services, that is, they do things for people, not make them. We are concerned, then, with things made in factories or large works. There are many ways of classifying these activities.

Perhaps the first group should be the metal-working industries, because much other industry depends on their products. The refining into pure metal of ore which is dug from the ground is called smelting. The most important industry of all in our modern world is the iron and steel industry. The shaping of pieces of steel or any other metal into wheels, machines, engines and indeed almost any other pieces of apparatus is the engineering industry. A textile is any fibre that can be woven, and the making of cotton, wool, silk, flax and other substances into cloth is called the textile industry. Another large industry is the chemical industry, which uses sulphur, salt and nowadays coal and oil to produce such things as soap, paint, detergents, explosives, plastics and nylon. The last is in turn a raw material for the textile industry. In our modern civilisation a great deal of our food is not eaten fresh; flour-milling, fruit-canning, making corn flakes, ice-cream, pork pies and so on is the quite considerable food-processing industry. These are the main groups of industries, though there are many others which cannot be classified conveniently, such as the making of paper, pottery, tobacco and rubber.

There are some other terms which you should know. You will often meet the expression light industry, which refers to the many small goods made nowadays, often in very modern factories. Cameras, radios or toys are examples. By contrast, heavy industry explains itself, and means mainly the production of large steel girders and plates, or of ships or railway engines. A processing industry usually refers to food, but the preparation of rubber and

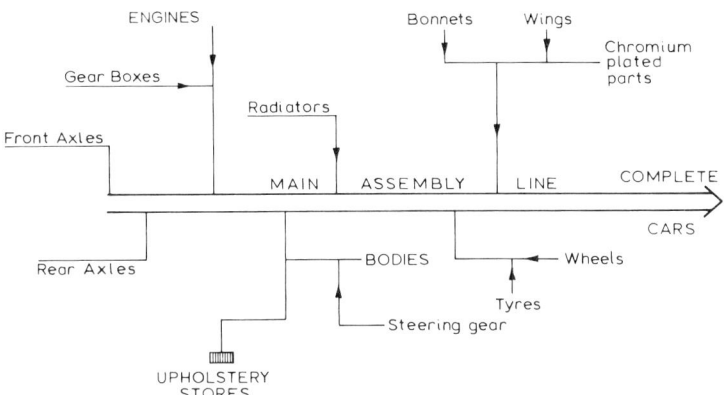

Fig. 4.1. A car assembly line.

power, raw materials, labour, i.e. workpeople, and site, i.e. a place to put the building or buildings.

Very often, as you will see, factories of the same kind or of different kinds are grouped together, in an industrial area. Their very existence also makes problems. From earliest times some industries have not been popular neighbours. There used to be tan-pits in Bermondsey which gave off an unpleasant smell, but fortunately they have now disappeared. A gasworks also sometimes smells, and it is not very pleasant to look at. Building houses near it is avoided if possible. You have already learnt

An industrial landscape: brickworks in Bedfordshire.
Refer again to page 23. What material is obtained from the pit on the left?
(*Aerofilms*)

tobacco could be included. Quite often you may hear of an assembling industry, where parts made elsewhere are put together; motor cars and radios are obvious examples (fig. 4.1).

The study of how industry is organised is part of the subject called economics. This would consider the problems of a man who wants to start a factory. He would have to find a suitable place to build it, and the money. He would need people to work in it, and power to drive the machines. He would have to be able to bring raw materials to his factory, and to distribute his finished product to the purchasers. Above all, he must be able to get all these things for less money than he receives for his produce, or he will go bankrupt. But factories, raw materials and power are found in different places the world over, which brings them within the study of geography, or, if you like, of economic geography. You will see later that it is not just a matter of getting a convenient place where all these things can be found, but of finding a place where the most expensive items to move have the shortest journey. We have to consider all the problems listed above if we want to know where industry grows up, and why. They are in summary

81

about air pollution. The cement-making industry gives off a great deal of white powder in the processing, and if you live near towns on the Medway or lower Thames you will have noticed the white dust coating the roofs, trees and gardens. There is a brickworks in Bedfordshire just east of the M1 motorway, along which you may travel. In spite of the tall chimneys, when the wind is from the east and atmospheric conditions cause smoke to come downwards again, you will certainly notice the smell. Modern factories also use a great deal of water, sometimes for cooling machinery, but also for washing and cleaning their materials. This must run out into a river, and is called industrial effluent. Unless carefully treated, and not always then, it fouls the river and kills the fish. Other kinds of industrial waste just have to be dumped, and often make blots on the landscape. The waste heap from a colliery is well known, and present methods insist on its being stored underground. The alkali industry in Cheshire also produces large quantities of useless white sludge, which must be dumped on waste land. Piles of broken crockery can be seen still in the Potteries, and near pulp mills in the great forested areas are often masses of waste timber and trimmings.

All this makes industrial landscapes often unpleasant and untidy. They are of many different kinds. Those which are most familiar in this country are in our coalfield areas, where industry grew quickly, before the days of planning. If you travel through Durham or Lancashire or industrial West Yorkshire you will see rows of houses, pit-head gear, tip-heaps, railway yards and factory chimneys spread over a large area, together with some of the original farms and farmland. Fig. 4.2 shows the features that develop in a mining and heavy manufacturing area. If you study the key and try to think what each item looks like you will realise how unattractive man has made this landscape. The use of hydro-electric power for metal-working in Alpine valleys makes another industrial landscape. On the flat floors of the larger valleys the beautiful mountain scenery is disfigured by long rows of smelters and engineering works.

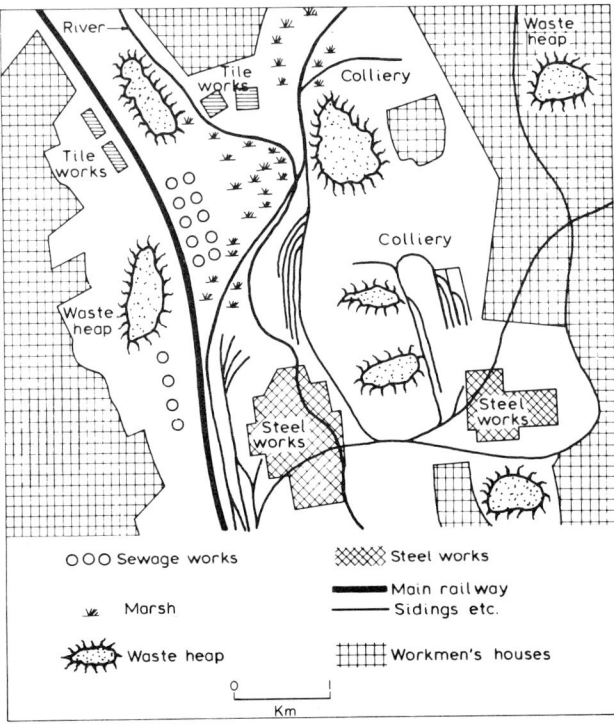

Fig. 4.2. An industrial landscape: the Potteries.

A large works: oil refinery at Shell Haven, Essex.
Note that this is a landscape entirely made by man.
(*Shell*)

The new industrial landscapes are cleaner and neater. There are some vast industries which require large works which cannot be hidden or beautified; steel-works, oil-refineries, chemical and cement works are such. Most others are housed in pleasant modern factory buildings. The centre of Eindhoven, in the Netherlands, consists mainly of the great Philips electrical works, but you would hardly see any difference from any other modern block of buildings. Sometimes the new factories are grouped together in areas called trading estates or industrial estates. These new industrial areas are cleaner, and have much freer choice of site, largely because they are powered by electricity. We must consider next the whole question of power supplies.

Work to do

1. *The table lists some primitive industries. Copy it.*

People	Simple industry	To make (products)
Indians in Equador and Yucatan	*Weave together the leaves of the screw-pine*	*Panama hats*
Bedouin of the Sahara	*Weave camel and goat hair cloth*	
Eskimos	*Sew bear and seals skins together*	
Amazonian Indians	*Hollow out tree trunks*	
Indians in Mexico and South America	*Use local clay*	
Africans in Ghana and Nigeria	*Drain the sap from a type of palm*	
Greek islanders in remote islands	*Spin wool from their sheep*	

i) *Complete the table. Add any other examples you think of.*
ii) *In what ways do these industries differ from modern industries?*
iii) *Are they similar in any way?*

2. *Make a list of the different groups or types of industry, giving examples of products from each group.*

3. *Make up a diagram to illustrate what is written about the chemical industry (page 80).*

4. *Copy the diagram showing an assembly line for cars. Using page 96 to help you, add places from which some of the car parts come. Add any other sources known to you. Give the diagram a title.*

5. *Write a list of all the services you can think of.*

2. Power supplies

The possession of natural power makes a great difference to a country's wealth. Power comes either from a source of heat, such as coal, or directly, as from the wind. Indeed, we often list these sources together under the heading 'fuel and power'. They are needed for heating, lighting and various other tasks but we are here concerned with power for factories and industry. Again we must think all the time about the question of where power is available. Sometimes it can be moved easily, sometimes not. The development of our knowledge of electricity has revolutionised the problem of the location of industry.

Perhaps the simplest form of power available is the wind. Windmills were a very early form of power-driven machine. The richest countries would be the windiest, and the obvious site would be on a hill. Fast-running or falling water was another early source, and again wet and mountainous countries would be richer than dry countries of flat lowland. You have already studied this in more detail in Chapter 2. One special form of falling water, which was sometimes used in the past, and may be again, is the tide. If the rising and falling tide in an estuary can be controlled, a great deal of power is available. Long, narrow inlets, such as those on the coast of Brittany, often have a tidal rise and fall of 12 m or more, and the French are developing this source of power at Dinan. Another possibility for the future is the direct collection of the heat from the sun's rays by some form of reflector, and in this case the hot deserts, with their constant sunshine, would be the best location.

Wood and charcoal were used to produce heat for metal-working centuries ago, but a great deal of wood was needed. The development of the use of coal in iron-smelting was an important step in the Industrial Revolution, and the other important step was the invention of the steam engine which was best fired by coal. England played a leading part in these activities, partly because she possessed several coalfields. Other countries which had coal soon followed suit, and until the early part of this century the world's industries were mainly on or near coalfields. Great industrial areas grew up on our coalfields, and near those of the Ruhr in Germany and the Appalachians in the United States, particularly in Pennsylvania. Coal is a bulky material, not easy to transport, and in general it was cheapest to use the coal fairly near to the coal-mines. The factories were built near by, the raw materials were brought from all over the world, and as jobs were available, people began to move to these areas to live.

The invention of the petrol engine and diesel engine caused the next great change. A convenient date to remember is 1914. During the First World War petrol-driven cars and aeroplanes became important, and for the following fifty years there has been a steady decline in the importance of coal, and growth in the production of oil. Once an oil well has been sunk, the oil flows out under natural gas pressure (fig. 4.3). Large numbers of miners therefore are not needed, and in general oilfields do not become large centres of population.

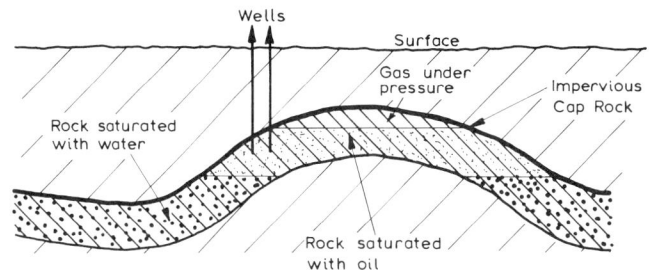

Fig. 4.3. Oil wells.

As oil is a liquid, it is relatively easy to move. It can be pumped through pipes on land, and is easily pumped into tankers for transport by sea. But much of it is not found where it is wanted. Countries such as Britain, Germany and the United States, which have large coalfields and many industries, are the prosperous countries which can buy oil for many purposes. Today one of the biggest world commodity movements is oil, moving from the Persian Gulf, the Sahara, Venezuela and the south-west United States to the highly industrialised parts of America and Europe (fig. 4.4). Oil production is itself another great modern industry, and the situation of refineries is another problem to consider. They are usually on the coast at the oil port, or at the end of the main pipeline leading to the consuming area.

The Second World War saw the development of atomic power, and that is the third great source. The basic material required is the mineral uranium, produced in the United States, the U.S.S.R.,

Oil depot: Billingham.
(*Aerofilms*)

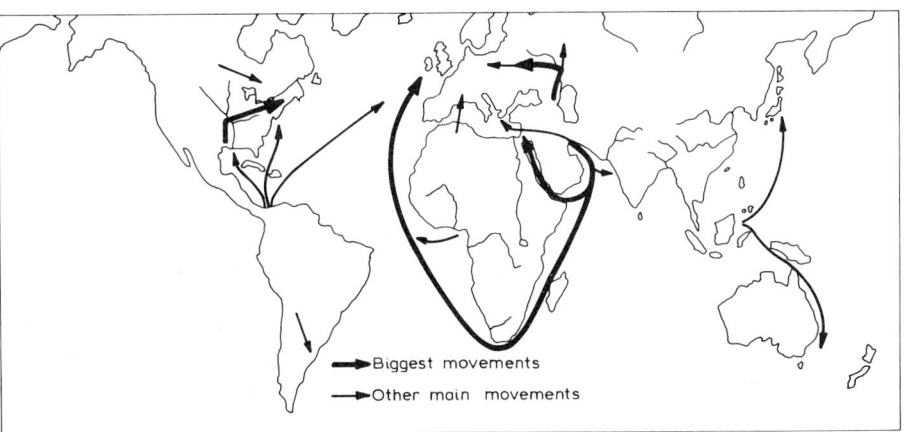

Fig. 4.4. World oil movements.

France, Gabon and Australia. As you know, a small quantity of uranium produces enormous amounts of nuclear power, so again transport is no problem. More important requirements are actually a large supply of highly trained scientists and large amounts of money to build the elaborate works needed to produce the power. Atomic power therefore can only be developed in technically advanced countries, such as the United States, Russia and western Europe. The other requirements are water and space. Atomic reactors need vast quantities of water for cooling, and are thus sited in this country on the sea coast. In the event of an accident they would be very dangerous, so the plants were at first set up in remote places. The early British reactors were in the far north of Scotland, and on the coast of Cumberland (fig. 4.5).

Now let us return to electricity. This is a means of moving

Atomic power station: North Wales.
(*Aerofilms*)

Fig. 4.5. Atomic power stations: England and Wales.

Fig. 4.6. Electricity grid.

power, and it is generated from turbines which are turned by falling water, or by engines driven by coal, or oil, or atomic power. Britain has an electricity grid which distributes power all over the country (fig. 4.6). The electricity is fed into the grid from hydro-electric stations in Scotland, from many thermo-electric, i.e. coal-fired stations in the Midlands and London, and from atomic power stations on the coast. If factories require oil, this, too, is easily distributed by pipelines. Another source of power also moved by pipeline is the natural gas now being sought beneath the bed of the North Sea.

All this means that for the last half-century in Britain and the other established industrial countries of the world, industries have not needed so much to be on coalfields. Manufacturers are more free to choose their sites, and to think of other problems mentioned on page 81. It also means that as coal is not used so much, and as the coalfields are not necessarily the best industrial areas, a considerable problem of unemployment has arisen. In Britain these areas of unemployment used to be called 'depressed' areas; now they are called 'development' areas, and the Government is trying to persuade industrialists to set up new factories in them to provide work.

The table below shows the electricity produced (millions of kilowatt-hours per annum) by the leading countries. The figures in brackets show that from water-power. You can therefore see the countries which make most of their electricity from coal.

United States	1638 (250)
U.S.S.R.	740 (124)
Japan	359 (80)
United Kingdom	248 (5)
West Germany	242 (17)
Canada	203 (156)
France	140 (56)

Work to do

1. i) *What was the earliest form of power-driven machine?*
   ii) *Name a second early form of power-driven machine.*
   iii) *What is the most commonly used form of power today?*
   iv) *What form of water-power may be used in the future?*
   v) *What vast form of power is as yet almost unused?*

2. *The table below shows the 6 major coal-producing countries of the world (other coal-producing countries mine less than 100 million tonnes a year):*

Country	Million tonnes of coal	Country	Million tonnes of coal
China	360	United Kingdom	145
Germany (West)	111	United States	541
Poland	140	U.S.S.R.	433

   i) *Rewrite the list in order of importance.*
   ii) *Draw, on graph paper, a column graph to show these figures, using a vertical scale of 100 million tonnes to 2 cm.*

3. *Copy the diagram of the oil wells (fig. 4.3). Write a paragraph to explain the diagram.*

4. *Draw a circle radius 6 cm to represent the total electricity produced in the seven countries shown on the table. From the centre of the circle measure 20° for Canada, 14° for France, 24° for West Germany, 36° for Japan, 74° for the U.S.S.R., 27° for the United Kingdom, and 165° for the United States. Colour the segments and key for each country. Give your circle a title.*

5. i) *What mineral is required for atomic power?*
   ii) *Name the countries in which the mineral is found.*
   iii) *Name two important requirements for the production of nuclear power.*
   iv) *Using the map (fig. 4.5), make separate lists of atomic-reactor centres in England and Wales.*

3. Raw materials and labour

Let us now review the many different raw materials which are used in factories. They are all provided by nature, and perhaps the biggest group is the minerals which are dug out of the ground. The metals, such as iron, copper, tin, aluminium and many others, are the most important, and the first process is to refine the ores in which they occur naturally. There are also non-metallic substances which seem less interesting and are more widely found. Clay and limestone, for instance, can be quarried easily. Salt and sulphur are obtained from the ground and are the basis of the chemical industry. Together with wood, coal and oil, they are also the raw materials for plastics or synthetics.

Some raw material is collected as it grows in its natural state, and timber is the chief example. Well-managed natural forests are nowadays replanted as they are cut, and it is a short step from this to their cultivation as commercial plants. Rubber is a good example. At first it was gathered wild from the Amazon forests, but now it is cultivated on plantations in other equatorial lands (page 174). Other crops grown for raw material, sometimes on plantations, are cotton, tobacco, jute, sugar-cane, oil palm and flax. The latter can produce a fibre for textiles or linseed oil from its seeds for food or lubrication. The rest of our raw materials come from animals reared by man. You will have thought of wool from sheep and some other animals, and hides for leather from cattle. Natural silk from the silkworm is also an animal product. Most animals produce more than meat to eat. Their bones, skins, horns and so on are usually processed in factories to produce fertilisers, glue and various other items.

You will realise by now that this book contains information about some of these things. You should look at the sections on minerals, forestry, plantations and farming. The last chapter on the world's regions will give you an idea of the kind of place from which most of the animal or vegetable raw materials come. We

Raw material: mechanical cotton picker: Mississippi.
For what industry is this a raw material?
(*U.S. Information Service*)

have drawn maps of where some of these materials are found or produced; these are called distribution maps. If you have an atlas which has maps showing where minerals or other raw materials are located you could draw your own distribution map from these.

In the past, as you have seen, raw materials were usually manufactured on the spot, but now transport is well developed and they can be moved readily all over the world. How far they move depends again on cost. Clearly small valuable items like diamonds cost very little to transport; heavy, low-value material like iron-ore or bauxite (the raw ore of aluminium) costs much more, although special bulk-carrying ships and loading apparatus at ports make this less nowadays. Some industries still occur near their source of raw materials. Fish preparation and fruit and vegetable canning are examples. Obviously it is better to have a canning factory in the growing area, so that the vegetables and fruits are canned while fresh, e.g. factories in King's Lynn preserve and can the vegetables and fruit of part of the Fenland and western Norfolk (fig. 4.7). In other cases the task is split up. Rubber is partly prepared in factories on the rubber estates in Malaya, and exported as sheet or crêpe rubber or even in liquid form. It is then made into finished goods nearer where these are wanted. Similarly, copper ore is partly processed in the mining

areas, such as the Katanga. To reduce the bulk some of the impurities are removed, and the resulting substance is called copper concentrate. But in many cases much raw material is moved long distances to places where industries already exist, and where people know a great deal about how to manufacture them into finished goods.

We can now consider the problem of the supply of labour. This is not just a matter of having large numbers of people to work in factories, as you will see. If this were so the main industrial areas would just be those with the densest population (fig. 7.1). It takes time for crafts and industry to develop, and the process of specialisation has been slowly growing throughout the history of civilisation. Once people have learnt how to make something, they will teach others. The medieval craftsman taught his skill to his apprentice, and so his knowledge was transmitted. These craftsmen formed a society, or guild, to share their knowledge. Again this happens now on a large scale. Once an industry is established in an area, there are many workers available who are skilled at the particular jobs needed. New manufacturers know, for instance, that if they set up a factory in Bradford or Huddersfield they will find there plenty of people who know how to do the work in a woollen mill. In the Midlands there are many skilled engineers working in motor-car factories. A further development is that local technical schools, knowing what jobs beginners want to learn, offer special courses in those subjects. Consequently, an established industry tends to stay in the same place, because the people of that district know how to work in it. Of course, you will also say that factories are not easy to move, and this is true, for plant is often large and machinery delicate. The tendency of industry to stay in the same place is known as industrial inertia, and it explains why some industries still exist although the original advantages of the area have long since disappeared. Sometimes you may read of the 'inherited skill' of the workers. This is not correct. The skill is not inherited, but the chance of learning it, if you live in such a place, is greater.

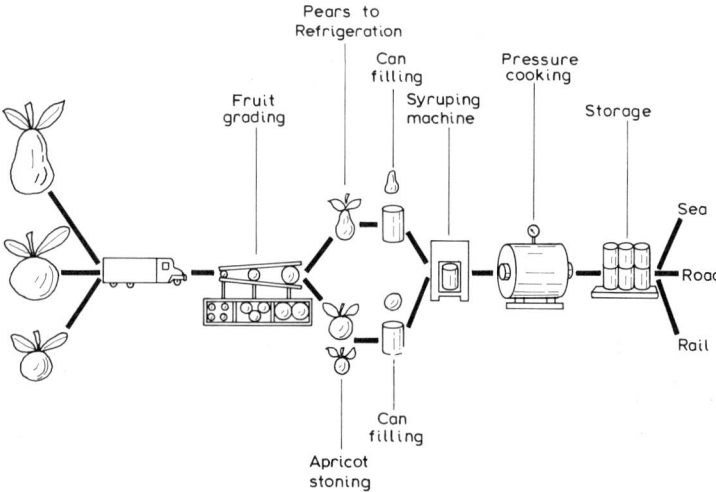

Fig. 4.7. Fruit canning.

The textile industry: spinning.
Note that one operator is responsible for many spindles.
(*J. Allan Cash*)

An interesting example of this is the Lancashire cotton industry. This grew up for various reasons, few of which exist now. For many years it continued to prosper, largely because there were many people who knew all about the spinning, weaving and selling of cotton. Even today, although the cotton industry has declined enormously, there are some new factories using nylon thread, so that the textile industry still continues.

This partly explains the problem we mentioned earlier. Not all the densely populated areas of the world are industrialised. In some of them, particularly western Europe, eastern North America, Japan and parts of Russia, industries have grown up for many reasons. In these places many people are accustomed to working in industries, and they know a great deal about it. In rural India, China, Africa and other lands the inhabitants are more used to working at other tasks.

Work to do

1. *The circle (fig. 4. 8) represents the 160 000 employed persons in Leicester.*

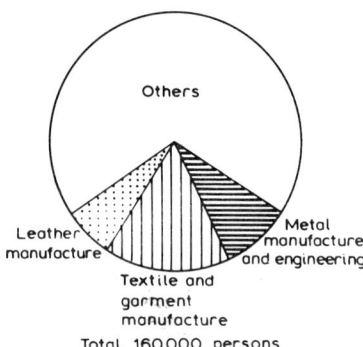

Fig. 4.8. Employed persons in Leicester.

i) *Work out how many people are employed in each of the three named industries. (Measure the angle of each segment at the centre of the circle. Put this number over 360 (degrees in a circle) to form a fraction. Find this fraction of 160 000.)*
ii) *How many people are employed in 'other' occupations?*
iii) *Make a list of ten occupations you know can be found in any large town.*

2. *Draw a diagram to show how all parts of an animal such as a cow or sheep are made use of (page 89).*

3. *Using the index of the book to help you find information, write a paragraph on bauxite.*

4. *If you wished to set up a woollen mill, where would you select a site? Give as many reasons as possible for your answer.*

5. *The table below shows Japan's imports in million dollars' value:*

Iron and steel	*150*	*Sugar*	*120*
Iron ore	*300*	*Raw cotton*	*520*
Iron and steel scrap	*390*	*Other metal ores*	*110*
Coal	*190*	*Wheat*	*180*
Soya beans	*120*	*Crude and heavy oil*	*700*
Raw wool	*350*	*Lumber*	*270*

i) *Rewrite the list in order of value.*
ii) *Underline industrial raw materials in red, food imports in blue.*
iii) *Which is probably Japan's largest industry? Which the second largest?*
iv) *Japan is short of raw materials for industry. Why is this?*

4. Sites and locations

Factories are set up at places where all the necessary items can be most cheaply assembled and the product distributed. The raw material and the power may come from far away. Water, if needed, and the labour force must be fairly close. All the necessities must be gathered together in the factory, on a spot suitable for building it. We tend to think automatically of coalfields as industrial areas, but this is no longer quite true, now that power can be more easily transmitted. Nowadays almost any area can become industrialised, providing it has power, good communications and, above all, a skilled labour supply.

Steel works in South Wales: Port Talbot.
Identify on fig. 4.9 the features shown in the picture. In what direction is the camera pointing?
(*Aerofilms*)

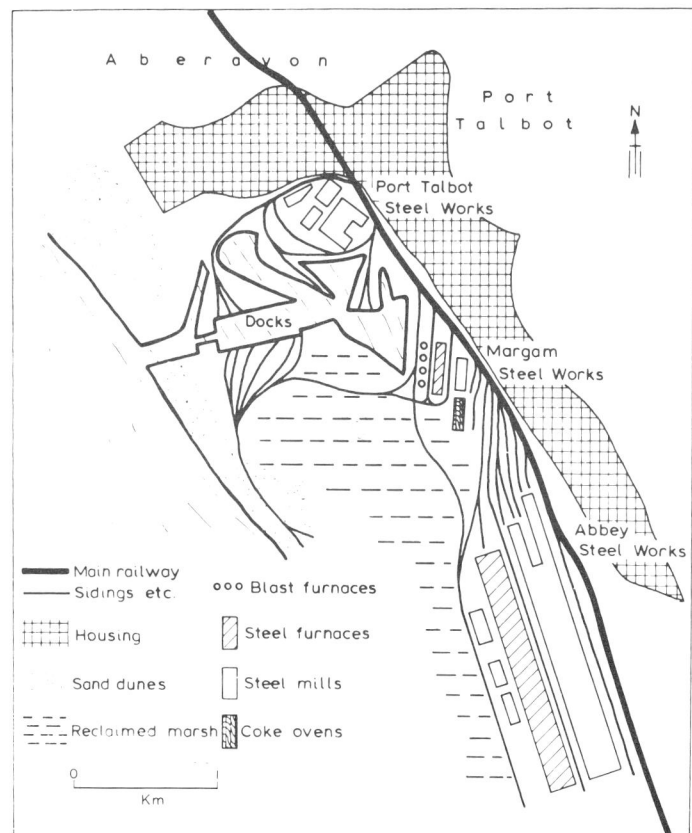

Fig. 4.9. Steel works in South Wales.

As our modern civilisation depends so much on steel, we will study first a large steel works in South Wales as an example of industrial location (fig. 4.9.). You probably know that first pig iron is made in a blast furnace. Into this iron ore, coke (which crushes less easily than coal) and limestone is placed and fired.

93

The resulting pig iron is then changed into steel in a converter or an open-hearth furnace. Both these require further great heat, and often scrap steel—another 'raw material' in this case—is added at this stage. The steel ingots produced are then shaped into strip steel or steel plates in rolling mills. They are then sent off to the many engineering works which need them as their raw material.

You can see that a great deal of space is needed, and in this case the waste sandy and marshy land was cheap and convenient. Ore comes from many different countries, and some coal moves by water, so a site by the sea, where a port can be made, is also important. Coal and limestone comes by train from other parts of South Wales, and scrap metal from almost anywhere, so many railway sidings are needed. These also take up much space. Steel-making does not need quite so many workers as some industries, but they must live somewhere, and many are found in the nearby town of Port Talbot. Finally, the product must be sent to the next factory: much moves by rail, and some by road or sea. The engineering works of South Wales, the Midlands and London use much of it. Large girders or ingots are awkward loads, but strip steel can be easily carried by lorry; you will notice such loads if you keep your eyes open for them. In general, steel works need to be near the industrial area they serve. British-made steel can be moved fairly easily by land to any part of industrial England. Large items of constructional steel, such as girders for overseas countries, have a short journey to the ports.

All this points to a site for a large steel works being on an expanse of flat land by deep water, where the bulky materials can be imported, and the extensive works built. Examples in other countries are at Sparrows Point, near Baltimore, on Chesapeake Bay, and at Gary, near Chicago on Lake Michigan. Even the older steel centres, such as those round Pittsburgh in Ohio and Essen in Germany, are on or near navigable waterways and good rail networks. Steel can best be produced where iron ore and power, usually coal, can be most cheaply brought together. The third ingredient, limestone, occurs very commonly and is seldom

Cars for scrap metal: New York.
What will happen to the cars before they become steel plates again?
(*Camera Press*)

a problem. The lowest-cost steel is probably produced at Birmingham, Alabama, where coal, ore and limestone are found within a few km of each other. The chief steel-producing region in Russia is in the Ukraine. Here the Donets coal area is linked by 300 km of railway to the Krivoi Rog iron fields. Wagons carrying coal in one direction can carry iron ore in the other, and steel works are found in both areas. The other important steel area has been developed in the southern Urals, partly for strategic reasons. This industrial area remained in production when other parts of the country were over-run by the invading Germans in the last great war. Now the coking coal comes from Kuznetsk, over

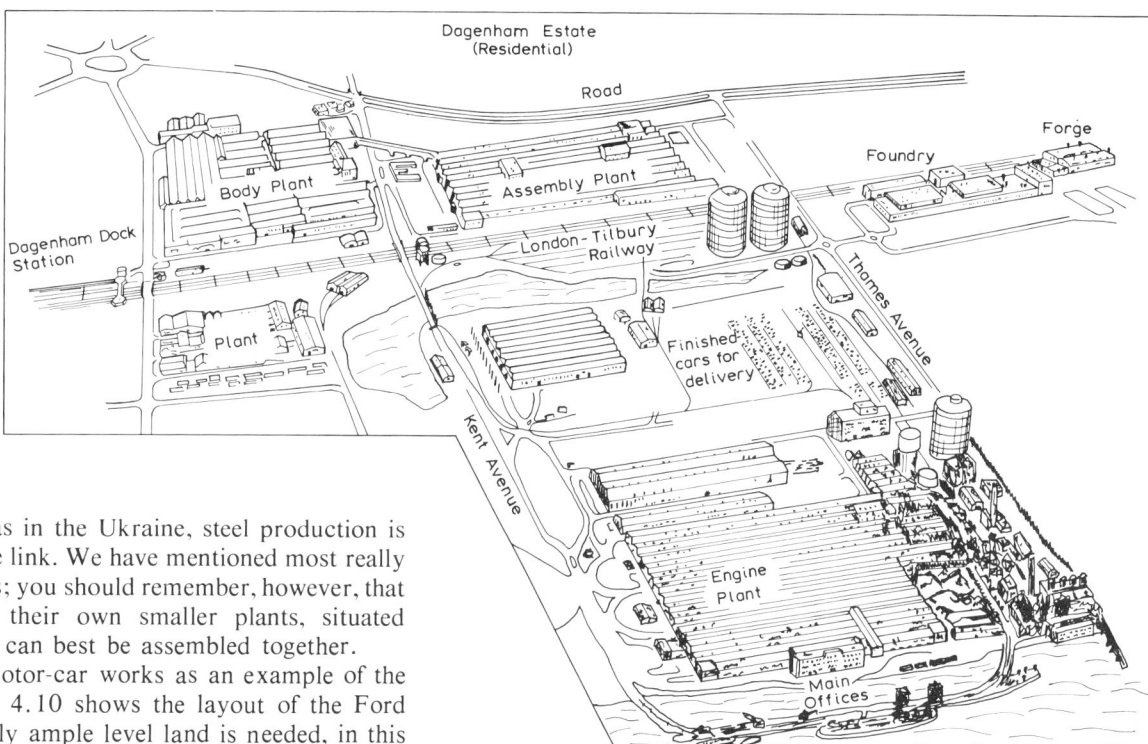

Fig. 4.10. Ford works at Dagenham.

1600 km to the east, and as in the Ukraine, steel production is carried on at each end of the link. We have mentioned most really large steel-producing centres; you should remember, however, that the other continents have their own smaller plants, situated wherever coal and iron ore can best be assembled together.

Now let us consider a motor-car works as an example of the engineering industry. Fig. 4.10 shows the layout of the Ford works at Dagenham. Clearly ample level land is needed, in this case reclaimed from Thames-side marsh. A main road and railway line are adjacent. Power is readily available, from the electricity grid and from water-borne coal delivered to the Thames-side wharf. The large Dagenham estate (page 113) houses many potential workmen. Labour supply is more important in an engineering industry, where there are many small parts to be made at individual machines. All the machinery must be under cover, and in the case of motor-car assembly, a great deal of floor space is needed. Important raw materials are pig iron and steel, which are cast in the foundry or shaped in the forge. There are many other

materials needed: paint, rubber, upholstery, wood, carpeting and so on. These mainly come by road. This is a highly organised firm, and a large office block is needed, as is space for the finished vehicles to be stored before they are sent off to dealers.

What is typical of the motor-car industry is that no longer is the whole of the manufacturing carried on at one site. Ford's now

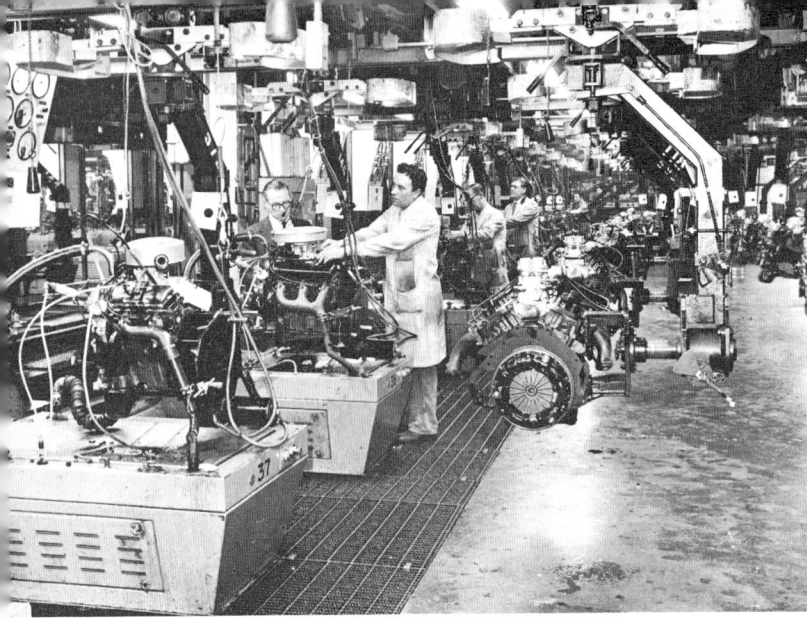

Engineering industry: Ford works at Dagenham.
Note that one man is working at each machine.
(*Ford Motor Company*)

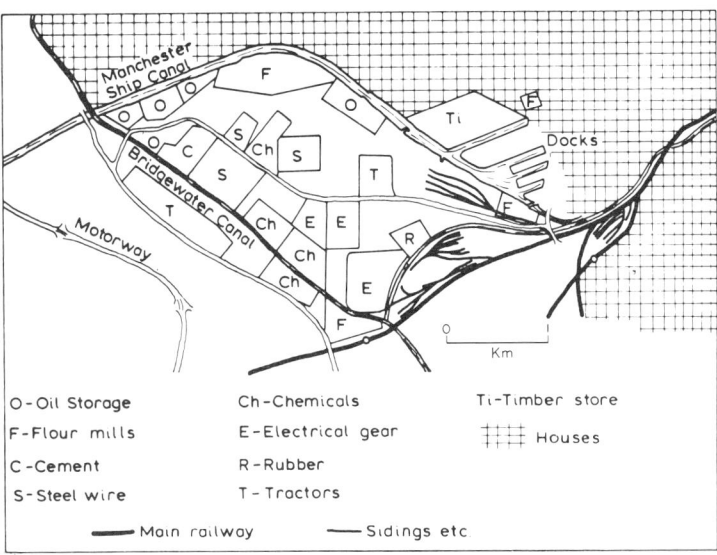

O – Oil Storage   Ch – Chemicals   Ti – Timber store
F – Flour mills   E – Electrical gear   ⊞ Houses
C – Cement   R – Rubber
S – Steel wire   T – Tractors

——— Main railway   ——— Sidings etc

Fig. 4.11. Land use at Trafford Park.

have many subsidiary factories. They acquired a motor-car body firm, with works at Doncaster, Southampton, Croydon and Romford. Radiators and other components are made at Basildon in Essex, and now that another large works has been set up in a very similar situation on Merseyside, at Halewood, all the gearboxes are made there. The pattern is the same in the other vehicle-making firms. Although there are main assembly works in Birmingham, Cowley, Coventry, Luton and other towns, component parts are made in small works all over the industrial Midlands and southern England. A major problem is that of organisation; the supply of parts from one factory to another must be carefully controlled, and a strike by the makers of an essential item may stop the whole process of production.

We have so far considered single large works, but you can see that circumstances which suit one firm may suit others. Fig. 4.11 shows the grouping of works in Trafford Park, on the west of Manchester. They all need bulky raw materials, such as oil, grain, timber, steel and rubber. These can be brought in by the Manchester Ship Canal, and there are also many railway yards. Similar concentrations of industry can be found also by the shores of the large estuaries of this country, such as the Mersey, the Humber and the Thames. Paper-mills and cement works are particularly noticeable on the Thames, and all have large collections of oil-storage tanks and flour-mills.

Another kind of industrial grouping, light industry, is often found on the outskirts of large cities, where smaller articles are

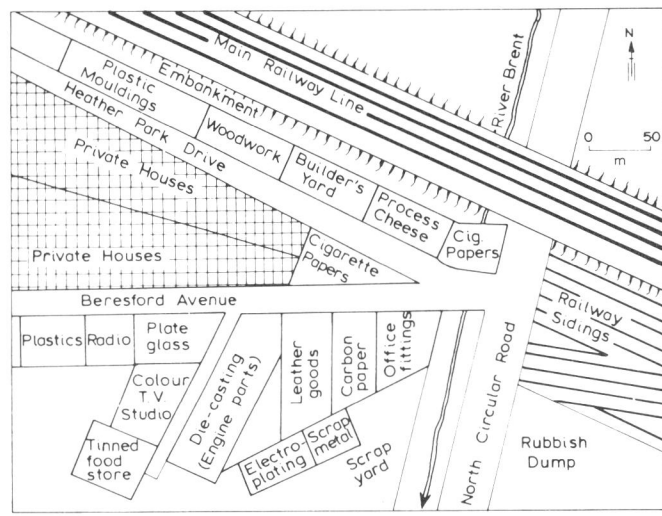

Fig. 4.12. Factories by the North Circular Road.

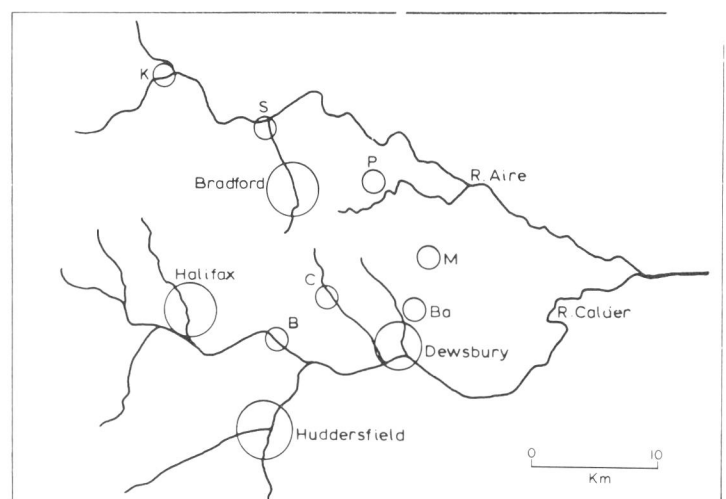

Fig. 4.13. West Yorkshire woollen towns.

made, which can be distributed by road, and which need smaller quantities of raw material. On the sample area (fig. 4.12) near the North Circular Road in north-west London you will find a hap-hazard mixture of products—office equipment, radios, paper and foodstuffs such as are constantly needed in the thousands of shops and offices of London. If you continued along the North Circular Road you would see more factories which make other goods of this kind, such as cosmetics, ice-cream, potato crisps, medicines and furniture. These are called consumer goods.

You may have learnt previously many details of other industries which we have not mentioned. Nowadays we think more of the general concentration of industry in certain areas, although of course there are still certain industries in special places. Shipyards could hardly be away from deep water, for example. The old

industrial locations can still be found in this country, but as we mentioned, their original advantages have nearly all disappeared. Our cotton mills are still in Lancashire, but they are no longer as important as in the past, and far more workers are employed in other industries in that county. In the group of towns in West Yorkshire shown in fig. 4.13 half the workers are occupied in the woollen mills, which still use the soft water from the Pennine streams.

These industrialised areas are found only in those parts of the world which have advanced technical development. We have drawn examples so far from this country, which is part of the highly industrialised region of north-west Europe. Parts of Russia are equally industrialised, chiefly the Moscow–Leningrad, southern Ukraine and southern Urals regions. The other great industrial area of the world is in the north-east of the United States and the adjacent parts of Canada. Japan has an equally high proportion of industry.

The other parts of the world are less highly industrialised. China probably has the next largest amount, particularly in the middle and lower Yangtse valley. India's growing industries are mainly in the Calcutta and Bombay areas. In South America the nearest to what is called an industrial complex, that is an area of general industrial development, is in the great towns of south-east Brazil, São Paulo and Rio de Janeiro, and round Buenos Aires. In Africa the only industrial area of real importance is the Johannesburg district of South Africa, and in Australia in the neighbourhood of Sydney. This is, of course, a very short summary of world industry, and many towns in all these countries have small or even large factories producing a variety of goods. In general, however, these parts of the world are agricultural, often relatively thinly peopled, and you will read of their problems in Chapter 7.

Work to do

1. i) *Trace the map showing Japan's main industrial towns (page 160).*
    ii) *Beneath the map list the industries of these towns (page 160).*
    iii) *With the aid of your atlas, state two features of position which are common to all these towns.*

2. *Name six geographical advantages which help the Ford industry at Dagenham.*

3. i) *Using your atlas, draw a sketch-map to show the position of the Manchester Ship Canal. (a) First draw a section of the coast from Preston to the mouth of the River Dee. (b) Mark in Liverpool and Birkenhead (you can write their names 'in the sea'). (c) Mark in the River Mersey. (d) Mark in and name Manchester. (e) Draw in the Manchester Ship Canal. Put a key for the canal symbol. Name the canal in the key. (f) Give your map a title.*
    ii) *Name five cargoes carried on the Canal.*
    iii) *Give a reason why these types of cargo are sent by canal.*

4. *Complete the table:*

Country	Location of main industrially developed areas
China	
India	
South America	
Africa	
Australia	

5. *Write a paragraph to describe and explain the requirements for a good factory site.*

Chapter 5

PROBLEMS OF TOWNS

1. Houses, hamlets and villages

Most civilised people, except nomads, live in some kind of fixed abode, their house. These houses are very different in different parts of the world, partly because different building materials are available. You have probably studied the grass huts of some Africans, or the log cabins of early settlers in North America. Nowadays most houses are made of stone or brick or steel and concrete, though it is surprising how common the wooden or 'frame' house still is in America and Russia. Very few houses are placed all by themselves. In pioneer lands a settler might build his house on an empty piece of land, but the great majority of settlements are grouped together in some form or other.

Most houses which are not in a village in England are farmhouses, placed so that the farmer lives in the middle of his farmland. The smallest cluster of houses is called a hamlet, and a bigger cluster a village. Anything bigger than this is called a town, but there are no rules to guide us. It is easy to speak of a large village or a small town, and there are now many towns with a million population, and several with several million. In Europe and North America more and more of the people now live in towns; between 70 and 80 per cent of them live in settlements with over 20 000 people. Study the chart below, which gives the picture for the whole world in 1950.

Size of settlement	Number of settlements	Total population in millions	Percentage of world population
Over 5000 people	27 600	717	30
Over 20 000 people	5500	507	21
Over 100 000 people	875	314	13
Over 500 000 people	133	158	7
Over 1 000 000 people	49	101	4

Ancient British camp: Beacon Hill.
Draw an annotated section across this camp site to show its defensive position.
(*Aerofilms*)

Outside towns, most houses are clustered together in hamlets or villages, and it is not at all easy to explain why these are sometimes big, sometimes small. Certainly people seem to like to live in groups for company, to help each other or to share common tasks. In the past they lived together for protection. You can still see the remains of ancient British camps on the chalk hills of southern England, often surrounded by a rampart and ditch. People also nearly always cluster together where there is water and food. In areas of rich soil there is often a series of villages, each surrounded by farmland. The people can live together, yet go easily to their fields. If there is water available everywhere they might be evenly scattered (fig. 5.1), but in other places they might

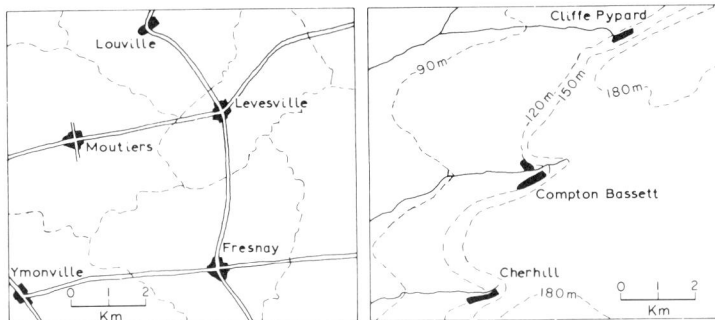

Fig. 5.1. Villages on rich farmland: Plateau of Beauce.

Fig. 5.2. Villages on a spring line: Wiltshire.

be dotted along a line of springs (fig. 5.2) or along a river. A suitable site for a village is therefore on level land, if possible, and near a water supply, but geographical circumstances are not always as simple as this. In mountainous lands villages will be on the least-steep parts; where floods are likely, the village will be on a knoll of rising ground. In very fertile lands, where all possible agricultural land is wanted, the houses may be squeezed closely together on a piece of rough ground, or on a hillside. History has also played a part. A village is often in the central part of a clearing made in the original forest, though nothing of the forest now remains, or by the walls of a medieval castle.

The great majority of villages all over the world are in farming areas; the inhabitants are near their work or their food supply. Sometimes other occupations are available; you will immediately think of a fishing village on a coastal inlet or river bank, or a mining village, often built specially, where nearly all the men work in one mine. Once established, some villages grow larger through being in a good position for trade or industry. They are then on the way to becoming towns, which we will consider in the next section.

Avegno di Fuori is an example of an Alpine village. It lies in a narrow, steep-sided part of the Maggia valley in Canton Ticino. The Maggia river flows into Lake Maggiore at Locarno. Avegno di Fuori means Avegno 'upstairs'; it has this name because it lies on an alluvial fan at the foot of a steep slope and is thus higher than its sister village of Avegno (figs. 5.3 and 5.4). It is south facing, so enjoys as much sunshine as possible, and lies a safe distance from the possible flooding of a little stream which flows across the alluvial fan.

The population of the village numbers 120 persons. Because it is situated near Locarno, to which it is connected by main road and railway, Avegno di Fuori is a flourishing tourist village where many householders accommodate visitors. It has a church, a café and some shops, but the larger shops and the local school lie in nearby Avegno. Originally most of the villagers were farmers, and

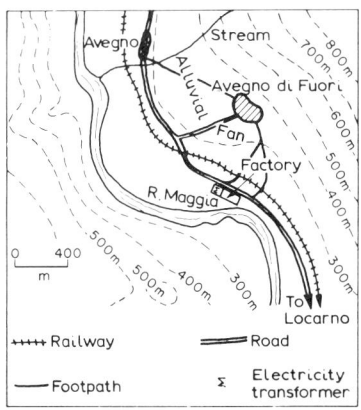

Fig. 5.3. The site of Avegno di Fuori.

101

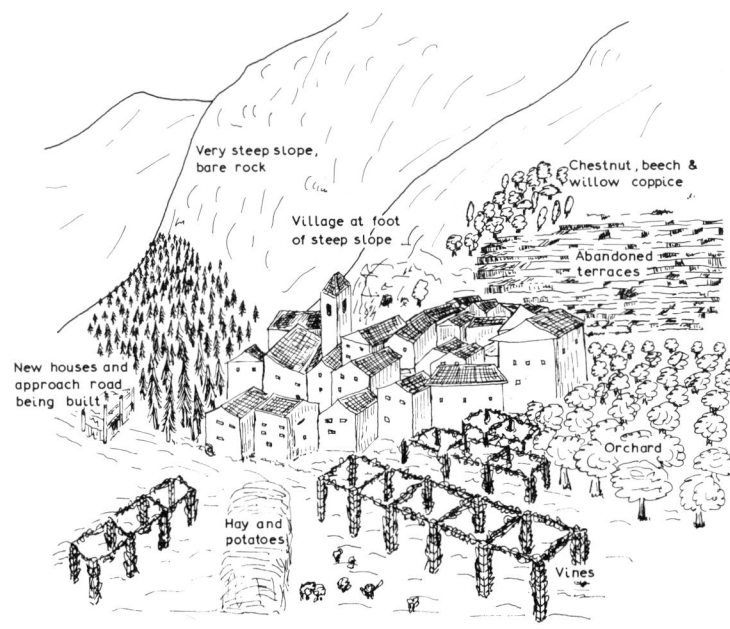

Fig. 5.4. Avegno di Fuori: a field sketch.

even now most of the women work on small plots of land to produce vegetables, salads and some flowers. The few farmers who still keep cattle send their small herds to the 'alp' on the high mountainside for summer pasture, along with sheep and a few goats. In earlier days the women used to spin the wool from their sheep and weave it, but nowadays it is easier to buy material in Locarno, where most of the villagers do their main shopping. Although vines are still grown on south-facing terraces, some of the terraces have been abandoned and wine production is now limited, for the men now go to work either in the nearby factory, which produces ball-bearings and other small metal articles, or in the silk, concrete-pipe or leather works of Locarno.

Work to do

1. i) *Construct a circle diagram to show the percentage of world population living in variously sized settlements (table, page 100).*
   (a) *Draw a circle radius 6 cm. This will represent the total world population living in settlements of 5000 people and over.*
   (b) *From the centre measure 144° for settlements of over 5000 people, 100° for those over 20 000, 62° for those over 100 000, 34° for those over 500 000 and 20° for those of over 1 million people. Colour the segments and put a key.* (c) *Give your diagram a title.*
   ii) *Do more people live in small or large settlements?*

2. i) *Give two reasons why people live in houses placed near to each other.*
   ii) *Where are villages found in mountainous countries?*
   iii) *Above what height are no villages found in the Alps?* (*page 152*).
   iv) *Why are villages sometimes found on a knoll of high ground?*

3. *Study the map* (fig. 5.3) *and answer the questions:*
   i) *What is the name of the highest village?*
   ii) *At what altitude does it lie?*
   iii) *Why is it not on the bank of the stream?*
   iv) *To what town does the road going south lead?*
   v) *Find this town on an atlas map and write a sentence to describe its position.*

4. '*In the 1880s the hamlet consisted of about 30 cottages and an inn, not built in rows, but dotted down anywhere within a more or less circular group. A deeply rutted cart track surrounded the whole, and separate houses or groups of houses were connected by a network of pathways. . . . From the hamlet a road led on the one hand to church and school, and on the other to the main road and so to the market town where Saturday shopping was done. It brought little traffic past the hamlet; an occasional farm wagon . . . a farmer on horseback or in his gig . . . and a carriage with gentry out paying calls in the afternoon were about the sum of it. No motors, no buses and only one of the old penny-farthing high bicycles at rare intervals. People still rushed to their cottage doors to see one of the latter come past.*'
   i) *What is a hamlet?*
   ii) *How big is this hamlet?*
   iii) *Write a paragraph describing the differences you would see in the hamlet today.*

5. *Write down a list of sites where villages commonly arise.*

2. Towns

What do you think of when you hear the word London? The Tower? London Bridge? The Thames? The Houses of Parliament? St. Paul's Cathedral? The City? Or going shopping in Oxford Street or to a theatre in the West End? Or just rows and rows of busy streets, crowded with traffic, or more quiet ones in the suburbs? Many modern towns are so big that they have many different activities, and can even be divided into very different parts.

Let us try to describe London using this idea. It covers an area of over 1500 square km, and has about 8 million people, so there is room for differences. What we now call the City is only 2·6 square km, and is almost exactly the area enclosed within the Roman wall, and later the medieval walls. Almost every building is now occupied by banks, insurance firms and the head offices of other large companies, and at night there is hardly a soul there except caretakers. Within walking distance is the River Thames and the port of London. The docks nearest to the Tower are now closed, and even those further downstream are losing their importance. If we go westwards we traverse Fleet Street and the newspaper district towards Westminster. Starting at Whitehall are more blocks of offices, but this time they are government departments, conveniently near the centre of government, the Houses of Parliament. Just north of Whitehall and Trafalgar Square is the West End, the heart of the entertainment area. There are nearly fifty theatres and large cinemas, not to mention restaurants and night clubs. This part merges into the busy shopping areas near Regent Street and Oxford Street, where there are the largest department stores, and some other shops, such as tailors and art-dealers, known all over the country, and indeed the world. We have not nearly described all the activities of this central area; there are many more offices, hotels, colleges, doctors' consulting rooms as in Harley Street, lawyers' chambers in the Temple and quite a lot of small workshops.

This busy part of a town is called the central business district, and occupies about 25 square km in the case of London (fig. 5.5). There is still much more. Of course, a great deal of the rest of London is residential, changing from close-packed houses and blocks of flats at first to more spacious houses in the outer parts. London is so big that there are within it many smaller 'towns', each with its shopping centres, offices and other services. There are also industries. Quite large areas of London are factory districts, in such places as the lowlands by the River Lea, among the docks and on Thames-side, or strung out along the main roads in the outskirts. Our list is not yet complete. What of airports, parks and football grounds, railway stations, waterworks and so on? You will be able to think of more activities and buildings which are characteristic of large towns.

We have seen something of what a large town is like, and most towns are, in many ways, similar. We can now think about what a town does, and this goes a long way towards explaining its growth. Most towns have activities which concern the people in villages and other settlements around them. London is so big that it serves not only its own 8 million people but much of south-east England as well. It even performs some services, such as insurance, for the whole world.

In the past, towns were built within a city wall for defence, though this does not happen today. Indeed, in an atomic age a city is an easy target, and a problem of defence is to get the population *away* from the town. Perhaps the chief cause of growth is commerce or business. Country people took their goods to market, and small towns grew up as meeting-points for trade. All the vast activities of the city of London, with its merchants, offices and banks, are really only an expansion of this idea. So are the large shops of any city centre. People go there because they know they are sure to find everything that they may want. If the town is a port, even more trade may develop, and we shall consider ports

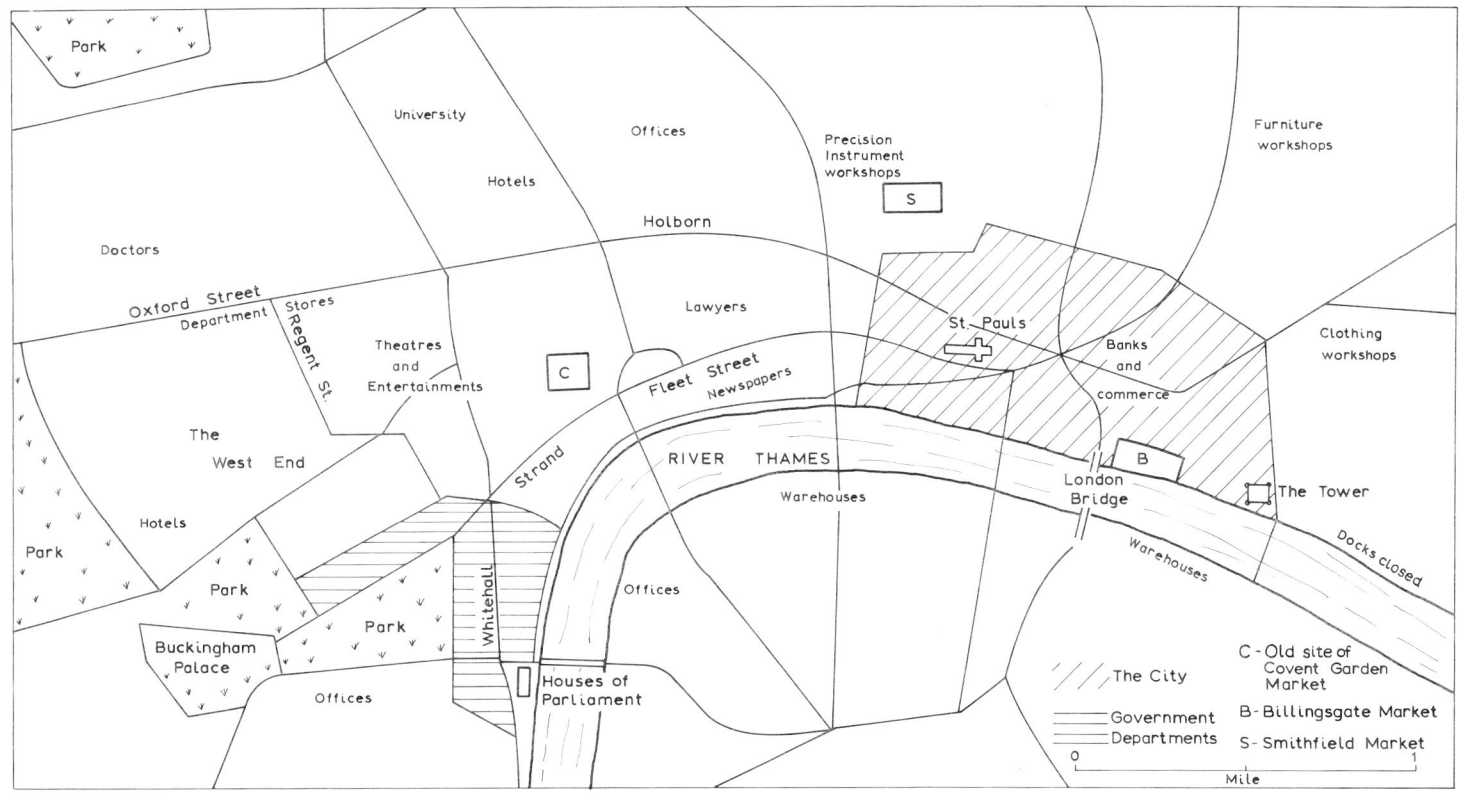

Fig. 5.5. Zones in Central London.

105

Walled city: Aigues-Mortes, France.
(*Aerofilms*)

later. We also saw that London is the centre of government, with many offices. So are many towns; there has to be a centre where the organisation of the district, for example, the running of schools, hospitals, the fire brigade and other services, is carried on. This is called administration. People also go to towns for entertainment and educational pursuits. Before the days of television the smallest villages had a bus service to take the inhabitants to the cinema in the nearest town, and many still have such a service. A large secondary school or hospital, for instance, is also usually situated in the town, to serve the district round it. Finally, we must think of industries. From earliest times craftsmen who could make things gathered in towns. Perhaps they knew that people coming to market would buy what they had made. As

industry developed, it was found that certain places had great advantages for making certain goods, and these became industrial towns. You have read more about this in Chapter 4.

Most towns have some or all of these activities, or functions, which have been described. The modern way of considering a town is to say that it is a service centre for the region around it. You have probably learnt about cotton towns, like Bolton, or woollen towns, like Bradford, but these adjectives are correct only when one activity in the town is clearly the most important. Is Oxford a University town or a car-manufacturing town? You can safely say that Blackpool is a holiday resort, Grimsby a fishing town or Pittsburgh a steel town, but most towns have many activities such as we have described. Above all, they are places where people have gathered together to live; that is, they are residential.

Site

The original site of a village or town may have been on firm ground at a river bank, on a well-drained patch of gravel among otherwise damp land or at some other suitable spot. In level areas with no great differences the first settlements may have grown up almost anywhere. Whatever the first reason, a town soon outgrows its original site, and if it expands any more people have to make the best they can of it.

There can be, therefore, problems of site. Obviously the first need is sheer space to grow. In general, the most populated parts of the world, and that is where most towns are, are on fairly level ground, so towns can usually spread easily. Modern towns need a great deal of space, however, for airports, railway yards, parks and so on. There is a limit to this sprawling outwards; as a result, there is increasing growth upwards. 10-, 20- and 30-storey blocks of flats are not uncommon nowadays. Some towns are awkwardly cramped by a natural barrier, often the marshland by a river, for example at Hamburg on the River Elbe. Rio de Janeiro, which has

The crowded city: central New York.
Using the pictures on pages 99 and 112 and this scene of New York,
list the common features of a large town which you would see from the
air.
(*U.S. Information Service*)

a splendid site in many ways, is blocked in places by sudden steep hills impossible to build on. The town of Caracas in Venezuela also expanded very suddenly, and the only space available was on steep and inconvenient hillsides. Water can also cause a problem. New York has many advantages of site, but it is spread over a number of islands and a river nearly 2 km wide, the Hudson. As a result, most railway lines end in New Jersey, and many ferries ply across the Hudson and between the islands. Tunnels and bridges are difficult to build, and all this increases New York's traffic problem (fig. 5.6). San Francisco and Stockholm have similar difficulties.

The actual ground can present problems too. London is lucky: the soft London clay underneath is easily tunnelled for the Underground railways, but the hard granite on which Rio de Janeiro is built makes tunnelling impossible. Soft, marshy land is obviously difficult to build on. Amsterdam requires special drainage and has few tall buildings. New Orleans, on the Mississippi delta, is also on land with a high water-table, that is, with water near the surface of the ground. Holes dug fill with water, and for many years there were no burials as such. Special cemeteries had to be built above the ground. There are many other towns where builders have faced these problems of site and underlying rock.

Health

When people are living close together in cities disease is easily spread, and often arises from the insanitary conditions of towns. A pure water supply is most important (page 40). Fresh water from springs or clear streams is seldom enough for large towns. The river water must be purified, and if necessary stored in reservoirs. In many new African, South American and Asian towns the inhabitants still have to take untreated water straight from the river. Rivers may not supply enough water; Manchester gets more by aqueduct from the Lake District, and Birmingham from Wales. Los Angeles, with its dry climate, has an aqueduct 240 km long from the Sierra Nevada mountains.

The waste water from houses and factories must next be carried away; it is called sewage. The sewers of a town normally run downhill, in drains under the streets, to a works on flat land by the

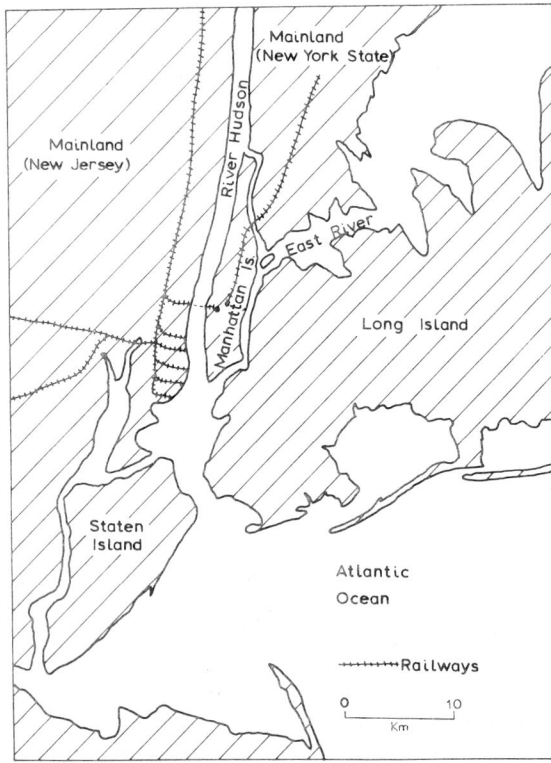

Fig. 5.6. New York: an island site.

river, where the sewage is treated so as not to pollute the river further. This system is expensive, and in plenty of towns there are only cesspits, covered deep holes into which house refuse runs.

Solid rubbish also presents a problem; there are enormous quantities to be carried away somehow. The most modern flats, have pulverisers which reduce rubbish almost to powder which can be flushed away, but this is rare. At the other end of the scale it is quite a common thing in tropical countries to see vultures eating garbage thrown on to waste land. Sometimes the waste is used as fertiliser; in China the need is so great that sewage is collected in carts and carried to the land outside. In most western cities household rubbish is collected by the dustman, but it must still be put somewhere. Sometimes it is burnt in an incinerator, or carried away to fill in old quarries or gravel pits. Seaside towns can have it loaded into barges and dumped out at sea, as does London.

Much smoke is also made in towns, from house and factory chimneys, and nowadays a great deal from car exhausts. This is the problem of air pollution, and modern towns are increasingly making laws against smoke (page 5). In the past air pollution has affected the shape of towns, for often the factory area has spread downwind, and the better residential quarters have been built on the other side of the town. This problem is partly overcome by the provision of open spaces; if there are plenty of parks the town is more healthy, and this is a question of town planning (page 114).

Supplies and markets

The people who live in towns need to be supplied with large quantities of fuel and food. This is the problem of distribution. Lighting and heating from coal makes the provision of large spaces for coal yards necessary. Nowadays, with more use of electricity and oil fuel, these are disappearing, but sites for large power stations and oil tanks must still be found, usually on the outskirts.

People also need fresh food, particularly milk, vegetables and meat. Around most very large towns is often found a market-gardening zone. The vegetables eaten in Tokyo come from a zone of market gardens concentrated in a circle of 50 km radius from the city centre, the most important vegetable producing area in Japan. The State of New Jersey has a similar large area to feed New York. Supplies of milk can be carried greater distances in tankers, but even so, there is often special milk production near a town, from stall-fed cattle. The home counties of England have a noticeably larger number of milk cattle than other parts of the south-east, though they are not rich grasslands.

You have already seen how often towns grew up as market places. In the centre of many small English towns can still be found an open-sided stone building, sometimes called the butter market. The centre of many towns is still called the market square. There are also towns in some countries, for instance, France, where country people still bring fresh farm produce to market themselves. In the bigger towns there are now specialised markets, where shopkeepers buy from the merchants who bring the goods there. Covent Garden, Billingsgate and Smithfield are well known as the specialised London markets for fruit and vegetables, fish and meat (fig. 5.5). In New York the specialised markets have been moved out from the crowded island of Manhattan to the outer parts; for example, the fruit market is now in the Bronx district. The Paris market called Les Halles has also been moved out from the centre. Sometimes the central part of a town is cleared and rebuilt, and special arrangements for a market are made. The new Bull Ring in Birmingham is such a planned market.

There are also other kinds of markets. The Oxford Street department stores form a market, specialising mainly in clothing. The City of London is a market, where insurance, shipping services and even money, in the form of loans, can be bought.

Finally, in each of the suburbs are local shopping centres, or even little groups of a few shops, which are markets for everyday things. Large and small, specialised or general, we find in towns many kinds of market.

The journey to work

The inhabitants of towns must earn their living; towns are places where people work. Poorer people must keep their fares down, and their houses are often near the factory areas; in old towns they may be mixed up among the factories. The larger the town, the more people who live on its outskirts. Most of the jobs are in the central business district, and those who travel regularly to and from the centre are called commuters (fig. 5.7). A million people travel daily to the centre of London; those who work in Manhattan, the centre of New York, spend an average of one hour on their journey. Electric trains are very useful for this kind of traffic, as they can call on enormous power for rapid starting after each stop.

The rush hour is a very well-known feature of large towns, when hundreds of passengers pour out from trains and buses (fig. 5.8). Handling so much traffic requires very careful organisation,

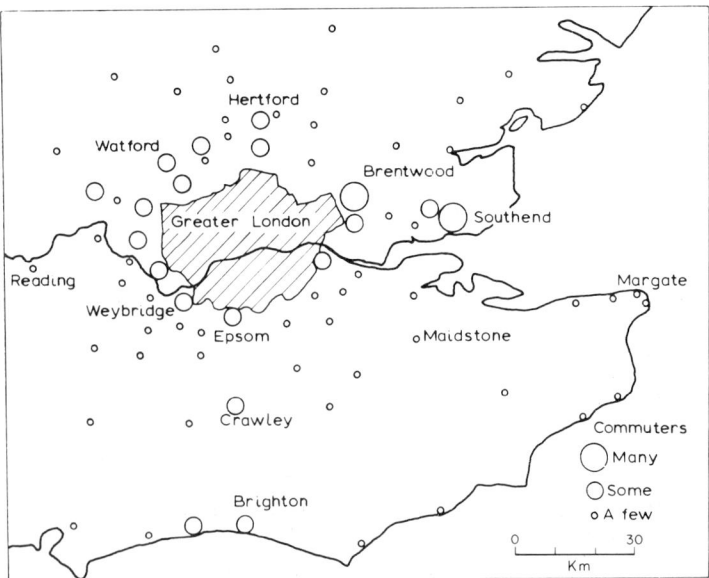

Fig. 5.7. Commuters: South-East England.

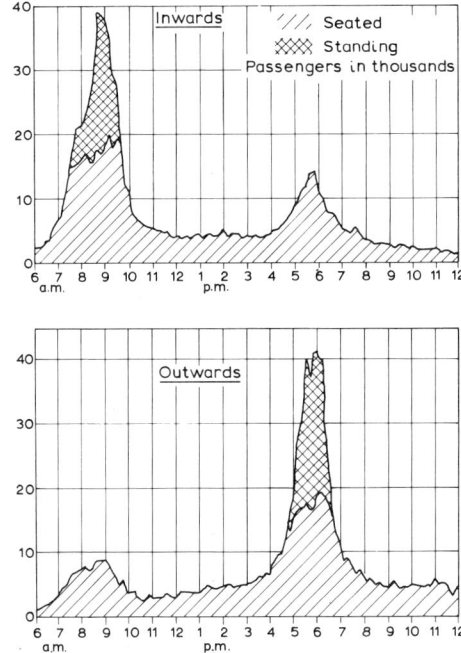

Fig. 5.8. London Underground rush hour traffic.

Fifth Avenue in the rush hour.
(*U.S. Information Service*)

Traffic problems

There are more traffic problems than the rush hour. Many towns grew up at cross-roads, and in small towns you can still see this central congested point. Old towns grew up before modern traffic, with narrow streets within the city wall. Today these are sometimes impassable, or an old gateway, as at Canterbury, may enforce a single line of traffic. The 'grid-iron' pattern of new cities avoids complicated cross-roads and permits better traffic flow. However good the roads, cities are now so busy that traffic congestion is a serious problem. It is very costly to widen or make new roads, as the owners of buildings knocked down must be compensated. Extra

and the bigger the town, the bigger the problem. The outer residential parts of the towns are called dormitory suburbs, that is, where people sleep. Everybody in America thinks of Westchester County as a residential area for New York, and in Britain of Surrey or Southend as the homes of city workers. This is more true of 'western' cities. In the great towns of Asia, such as Calcutta and Bombay, or in new African towns, such as Ibadan, there are not such clearly marked zones. Houses, shops and workshops are more usually mixed up together. There are few commuters, and a much sharper change from town to country.

City gate: Canterbury. ▶
(*Camera Press*)

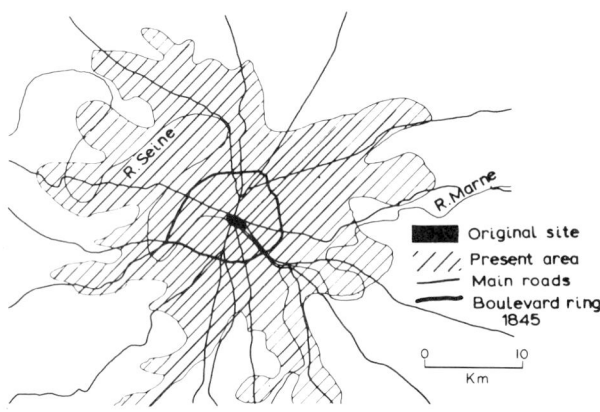

Fig. 5.9. An early ring road: Paris.

City centre: Paris.
The river is a barrier to communications in Paris and many bridges are needed.
(*Camera Press*)

space, already expensive, must also be found for car parks. One solution of the traffic problem is to prohibit cars from the centre of towns, another is to make more use of public transport. The Underground railway system is well developed in many large towns; some could hardly exist without one. The Moscow stations are very well designed, the Metro of Paris equally famous.

Many attempts to solve traffic congestion are tried. One-way streets, and traffic controls, such as no right-hand turns, are well known. You may have discovered some details of a local plan during your own traffic count. Some London bridges now use a 'tidal flow'. That is, in the morning there are three lanes inwards, and one outwards; in the evening the arrangement is reversed. The central area can also be avoided by a ring road. The earliest of ring roads were the boulevards of Paris and Brussels, though now they are far inside the built-up area (fig. 5.9). The last stage is the growth of elevated roads. Many American cities have various fly-overs, clover-leafs and other new road systems, built up like bridges above the other streets.

Growth and decay

Towns grow—mainly outwards. They spread along roads (ribbon development) or round railway stations. They spread over the best land, avoiding hills and marshes. The centre is the oldest part, and old buildings get pulled down. Some city centres have been rebuilt many times, though, as we have seen, the street pattern remains. Nowadays the centre has the finest new buildings, with just outside it the oldest and dirtiest houses which have decayed and are due for clearance (fig. 5.10). Slum clearance has long been an urban problem. The occupants must be looked after while new houses are being built. The new houses may cost more than they can afford, even to rent. Sometimes complete new estates are laid out. The London County Council estate at Dagenham was one of the biggest. Today whole new towns, called 'overspill' towns, 80 or more km away from the main one, are planned.

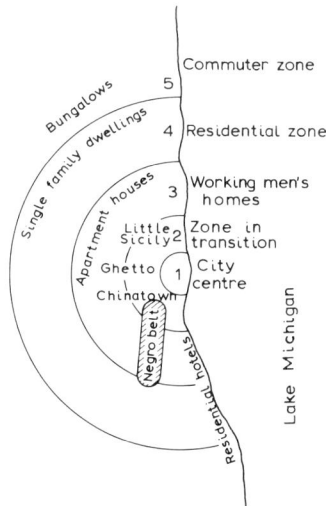

Fig. 5.10. Zones of growth: Chicago.

Urban reconstruction: Wandsworth.
List some of the problems caused by this reconstruction.
(*L. J. Long*)

In the 'young' countries of South America and Africa is another problem. People have moved rather suddenly from the country to the rapidly growing towns, before proper houses were built for them. These shanty towns are called 'bidonvilles' by the French—literally tin-can towns. They spring up almost overnight, on any patches of waste land, always on the outskirts. The one-room huts are made of scrap material, bits of timber, old metal sheeting or pieces of palm fibre. There are no roads, water or sanitation. The town of Caracas, growing rapidly through the prosperity of Venezuela from its oil, has 1 in 4 of its $1\frac{1}{2}$ millions in such conditions. Similar shanty towns have grown in Hong Kong,

113

built by Chinese who have fled from Communist China, and in Calcutta, to which Indians came from Bangladesh. The cost of rehousing the people in decent homes is very great, and proceeds very slowly.

Shanty town in Calcutta.
Describe these houses. What material are they made of? What facilities has your house which these have not?
(*Camera Press*)

Town planning

All these problems, and others, are the business of town planners. Previously towns grew almost haphazardly, without any conscious direction by man. Nowadays he tries to control their growth by a town-planning department, often employing geographers. By including architects and artists in the teams, he tries also to make them beautiful. Nearly all we have been considering in this section is the concern of town planners. They must judge where is the best area for new building; to do this they must think of water supply, drainage and so on. They must allow places for hospitals, schools and offices. They try to group the factories together in an industrial area, separate from the residential areas—which also need parks and open spaces. Each residential area needs its shops, and road and rail systems to these and to the factories. In old towns this is a difficult task, and town planning is sometimes just a patching-up operation, improving on the untidy sprawl as well as possible. It is easier to plan a whole new town from the start. The earliest such places in England were Letchworth and Welwyn Garden City (fig. 5.11). In recent years many such new towns have been built. In these a new problem is how to bring the inhabitants together. They are all strangers at first, and special plans must be made to get them to take an interest in their own new town.

Planned capital cities

A special kind of planned town is the artificially created capital city. The town which contains the central government of a country is called its capital. This is often the oldest or largest or most central of its big cities. In some cases, however, usually in federal countries, new capital cities have been planned. A federal country is a group of states, each of which has considerable independence, but which act together for certain important matters, such as

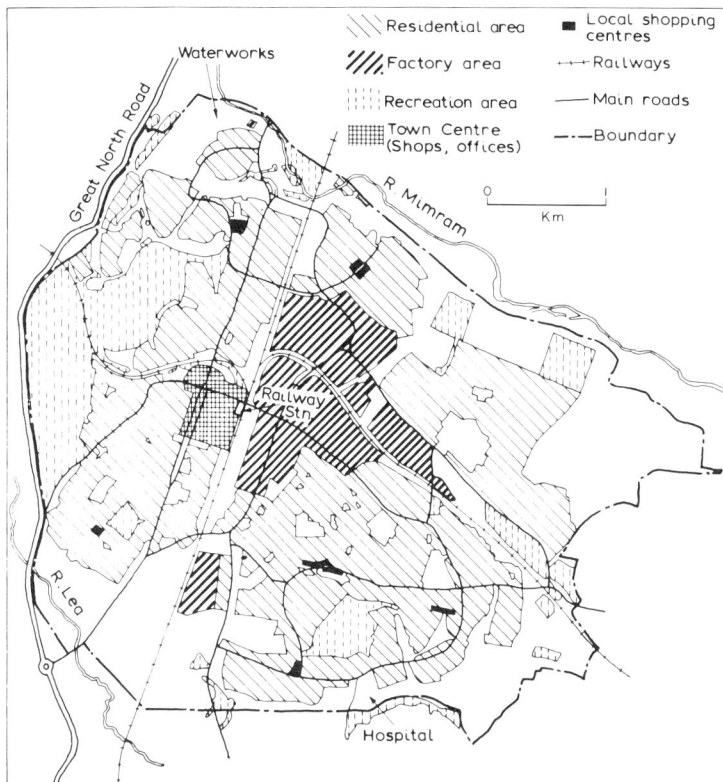

Fig. 5.11. Welwyn Garden City: a planned town.

A planned capital city: Canberra.
How can you tell that this is a planned city? Draw a map of the street layout.
(*Australian Information Bureau*) ▶

defence and other external affairs. The obvious example is the United States, but Australia and Brazil are also federations.

A piece of land by the River Potomac was set aside for the capital city of Washington, and called the District of Columbia. The capital is thus not in the territory of any state. It is a beautifully planned town, with provision made for the White House (the President's residence) and many government buildings. Industries were at first prohibited, but much printing is necessary in such a city, and there are plenty of hotels, as the town attracts many tourists. At the time it was established, in 1791, most of the population was on the eastern side of the continent, and Washington was then in a fairly central position. In the case of Australia, a situation had to be found which was near the chief areas of population, and the new capital of Canberra was planned

in the hills of the south-east, roughly midway between Melbourne and Sydney. The most recent capital is Brasilia, which has been deliberately placed in the interior of Brazil, nearly 1000 km north of the two main towns, São Paulo and Rio de Janeiro (fig. 5.12).

Fig. 5.12. The location of Brasilia.

Although elaborate new government buildings and elegant blocks of flats have been built in a very spacious plan, the capital is in a remote spot and has so far not attracted many inhabitants other than diplomats, civil servants and shopkeepers. The flats are expensive, and poorer employees are still living in cheap temporary huts outside the planned city. Many people who have to do business in Brasilia prefer to fly there for a brief visit from the old towns, and return home for the week-ends.

Ports

Ports are another specialised kind of town. The site plays an overwhelmingly important part in their growth, and as you know, they can also develop into great towns with all the other functions we have studied. What does a port need? Obviously a harbour, that is any water naturally sheltered from storms and waves. Almost any inlet or enclosed bay provides this. A place is also needed for ships to tie up and discharge cargo, that is, firm land by deep water. An area of flat land by the water is also needed to build the town on. Finally, there must be a district behind the port, easily accessible to it, which has goods it wishes to export or needs to import. This is called its hinterland.

You will notice that a river mouth or estuary fits these requirements, offering both shelter and a route inland. Often great ports are found on them, but they are not perfect. They may be shallow and muddy, or have high tides which make docks necessary. Marseilles is a good example of a port which avoids a shallow river mouth, that of the Rhone, though much of its trade moves along the Rhone valley. Long arms of the sea, such as rias or fiords, also offer good sites, but they may be set in rocky or stormy coasts, such as at Brest in Britanny, where navigation on entering is difficult and dangerous.

In the modern world ships are large, and most ports have manmade improvements. Some are completely artificial, for example, Manchester, dug out at the end of the Manchester Ship Canal, or Takoradi, which was built on the shallow surf-bound coast of Ghana. At most ports, however, the principles described above hold good: nature offers circumstances which help man. Jetties can be built to increase the natural flow of currents, which help to scour a deep channel. This can also be dredged. Wharves can be built with a sharp edge at deep water, and soft land excavated to build docks where ships can stay at all states of the tide. Much equipment is also built, such as marks for navigation, cranes, warehouses and railway sidings.

Most ports are for general cargo, but there are also highly specialised ones which grow in response to a particular need. The great iron-ore ports, such as at Duluth on Lake Superior, are little more than a series of jetties from which ore wagons can be emptied into the ships, and oil terminals, such as at Fawley or Milford Haven, consist of a jetty from which a ship discharges oil through a pipeline. The passenger ports on the Channel coast, with facilities for car-ferries, waiting rooms and so on, are another example.

Now study figs. 5.13 and 5.14, to see how far the port of Liverpool illustrates these ideas. You will find there is a narrow-necked estuary where currents are strong and water is deep, and man-made improvements by the Crosby Channel. The long line of docks have been gradually built out into the river. The town of Liverpool has spread out over the flat land of south-west Lancashire. Liverpool grew in importance as industrial Lancashire developed, but it now serves a much larger area, including Lancashire, much of Yorkshire, North Wales, Cheshire

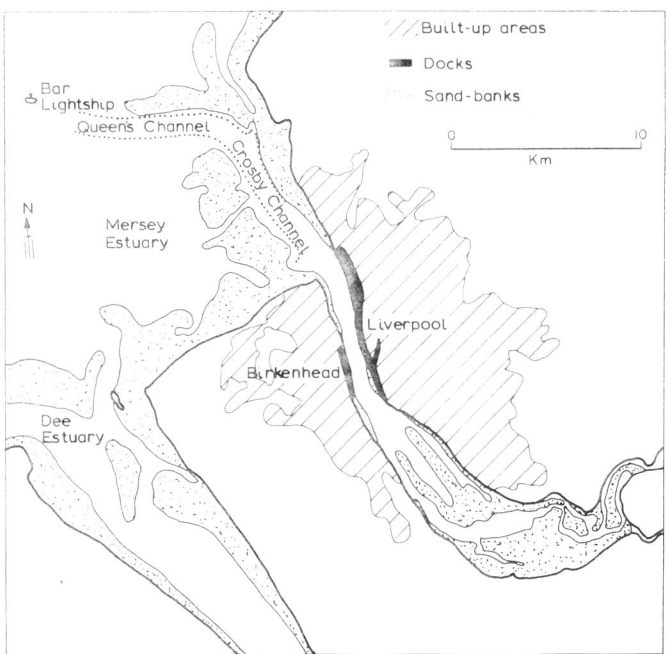

Fig. 5.13. The port of Liverpool.

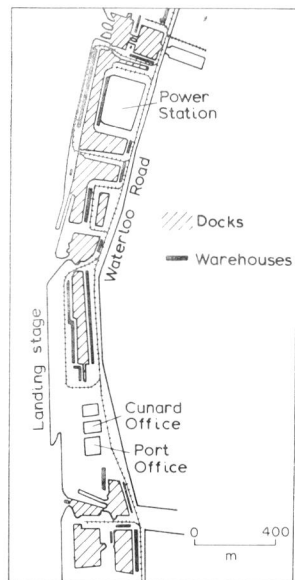

Fig. 5.14. Liverpool docks.

Liverpool docks.
(*Aerofilms*)

and the Midlands. If you mark these areas on a map you have drawn Liverpool's hinterland.

Work to do

1. *Study the map* (fig. 5.5) *and answer the questions:*

   i) *In which direction is the Thames flowing?*
   ii) *What is the curve of the river part of?*
   iii) *To which are the docks near—the source or mouth of the river?*
   iv) *Name all the types of buildings along:* (a) *the north bank;* (b) *the south bank.*

   v) *How many bridges are there? Name:* (a) *the bridge near the docks;* (b) *that leading to the Strand and Fleet Street;* (c) *that by the Houses of Parliament.*
   vi) *What buildings line Whitehall? Why is this?*
   vii) *Name the three markets shown and the produce they deal with.*
   viii) *Name the two main shopping streets shown.*
   ix) *Name the three industries shown.*
   x) *Write a sentence commenting on the position of the industrial areas.*

2. i) *Describe the ways in which London, as a capital city, differs from our other cities.*
   ii) *Do you think all capital cities are like London? Give reasons for your answer.*

3. *Give three examples each of the following types of towns from anywhere in the world:* (a) *holiday resort;* (b) *fishing port;* (c) *university town;* (d) *planned capital city;* (e) *steel town;* (f) *car manufacturing town;* (g) *coal mining town;* (h) *port.*

4. *Draw a sketch-map to show the position of any town named in Question 3. See that your map has at least four features marked which help locate the town clearly. Depending on the town, these might be a river, the junction of two rivers, a range of hills or mountains, a coastal stretch, a main road leading to a named town or any such feature.*

5. i) *Why has London an underground railway system?*
   ii) *Why are London's underground railways less profitable than formerly?*
   iii) *Give reasons why Amsterdam, New Orleans and Rio de Janeiro have no underground railways.*

6. i) *What is a commuter?*
   ii) *Why do people commute?*
   iii) *What are:* (a) *the advantages;* (b) *the disadvantages of commuting?*

118

7. *Study the map* (fig. 5.7) *and answer the questions:*

   i) *Name the two largest commuter towns north of London.*

   ii) *Name the three most important commuter towns on the south coast.*

   iii) *How far are* (a) *Reading,* (b) *Crawley,* (c) *Margate from central London?*

   iv) *Where are the greatest number of towns with many commuters in relation to London? Why is this?*

8. i) *What is a market?*

   ii) *What is meant by a specialised market?*

   iii) *Name the countries from which the following products are sent to Covent Garden market:* (a) *oranges;* (b) *apples;* (c) *walnuts;* (d) *melons;* (e) *grape-fruit;* (f) *early vegetables;* (g) *bananas.*

   iv) *Name six other products available (always or in season) at Covent Garden market.*

   v) *Name any town you know to have a cattle market.*

   vi) *Name three towns in the British Isles with large fish markets.*

9. *'Below, the city lay before them; the rain had stopped and a sharp wind was driving the clouds apart; great shafts of sunlight slanted down. Each in its accustomed place the Capitol, the White House, the Library of Congress, the Court, the Washington, Lincoln and Jefferson memorials, the medieval spires of Georgetown University, and the bulk of the Cathedral stood out. The river wound brown and muddy under its bridges, stretching away south and east toward the Chesapeake Bay;*

over the rolling countryside of Maryland and west along the Virginia Hills to the Blue Ridge and the Shenandoah the first light carpetings of green were beginning to show . . .'

   i) *Of what city is this a description?*

   ii) *What is:* (a) *the Capitol;* (b) *the White House;* (c) *Congress?*

   iii) *Why are there memorials to Washington, Lincoln and Jefferson?*

   iv) *Georgetown is the western district of the city. What building lies there?*

   v) *What is the name of the 'brown and muddy' river? Use your atlas.*

   vi) *Locate Maryland and Virginia in your atlas. What are they? What is Virginia famous for?*

   vii) *Where and what are:* (a) *the Blue Ridge;* (b) *the Shenandoah?*

   viii) *In what season does 'the first light carpeting of green' occur?*

10. *Two of Liverpool's many imports are oil and cotton (cotton is ninth in importance now). Oil comes from the Caribbean, Kuwait, Bahrein, Louisiana, Texas, Trinidad. Cotton comes from the Sudan, Natal, Bombay, Queensland, Turkey, Peru, south China, south United States and Egypt.*

   On an outline map of the world, as accurately as possible

   i) *Mark in and name the areas sending oil to Liverpool.*

   ii) *Mark in and name the areas sending cotton.*

   iii) *Mark in the main shipping routes.*

   iv) *Put a key and give your map a title.*

3. A world view

Although towns have existed since earliest times, the last century has seen quite a remarkable growth, as the following table shows:

World Population Living in Towns of over 5000 People		
1800	27·4 millions	3%
1850	74·9 millions	$6\frac{1}{2}$%
1900	218·7 millions	$13\frac{1}{2}$%
1950	716·7 millions	30%

This means that nearly one-third of the whole population of the world lives in such towns, and also that there has been an increasing dislike of living in the country. One example will suffice to illustrate this latter statement—the departure of large numbers of crofters and other people from the Highlands of Scotland, known as the rural depopulation of the area. Probably the chief cause of rural depopulation is the great growth of manufacturing industry in the last two centuries, known as the Industrial Revolution (page 80). Factories needed a supply of workpeople, and these preferred to live nearby the factories, but there were also many other reasons for this movement to towns.

The recent past has seen the growth of the very large town. A convenient size to remember is a population of 1 million, and such towns have been called 'millionaire' cities. Towns of about this size are Birmingham, Glasgow and Liverpool. In 1900 there were only eleven towns as large as this in the world, but today there are over 100. Nearly one person in fifteen lives in a town of this size or larger. These figures are obtained from the census figures (page 152) of those who live inside the town's official boundary, but towns grow rapidly, and there are often many new houses just outside this. The inhabitants, for all practical purposes, dwell in the town. This often untidy spread of a town is called 'urban sprawl'. Sometimes a group of towns grow so that they almost merge together into a continuous built-up area; this is called a 'conurbation' or 'urban agglomeration'. The industrial area of the Ruhr coalfield in Germany extends almost continuously from Duisburg on the Rhine for nearly 50 km eastwards to Dortmund, and contains about 10 million people.

Let us next see where there are most towns. You might think that they would be where there is most population, but this is not entirely true. We have seen that big towns have grown up mainly where much industry has developed. The most advanced industrial parts of the world are Europe and the eastern part of the United States, and as the map shows (fig. 5.15), more millionaire cities are found here than anywhere else. Japan is also highly industrial, and has many large towns. The other densely peopled areas are India and China, but here the situation is slightly different. The majority of the 1400 million population are still peasants (page 66), that is they live as farmers on their small plots of land, though in China they are now grouped together on large communally owned ones. We say that the population is mainly rural. Although most of the people are country folk, there are now some dozen really large towns in the Indian peninsula, and about fifteen or more in China. This is largely due to the growth of industry in those countries. Many of the people who live in them are still countrymen at heart, and this is another problem. They still have their country ways, and are not yet used to the very different conditions of life in towns.

There is a different situation in each of the southern continents. Australia is still a rather empty land, and as most of the immigrants are from Europe, they are townsfolk. About half Australia's 13 millions live in the five main towns. You could draw a map to show these; your atlas will tell you which they are. Africans in the main are agriculturalists who live a country life. There are some growing towns, but they are in fact more like overgrown villages. The only exceptions are the ancient cities of Alexandria and Cairo, and the urban agglomeration which has

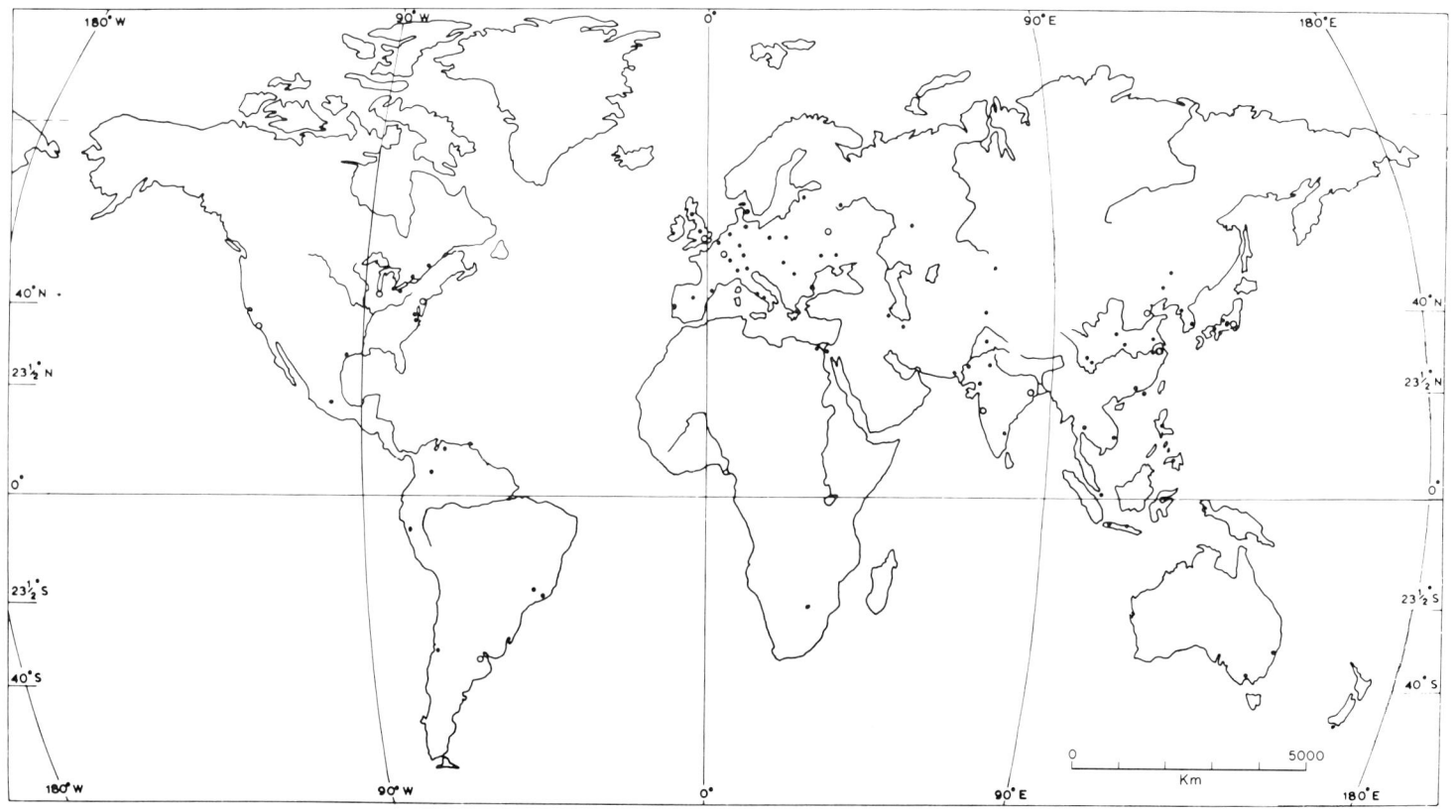

Fig. 5.15. World 'millionaire' cities. ● = cities of 1 million ○ = super-cities (over 6 million).

grown up around the goldfields of South Africa, centred on Johannesburg (page 25). South America is again different. Perhaps because its present peoples come largely from Europe, and are accustomed to town life, and partly because of its growing prosperity, there are several really large towns, such as Buenos Aires, Rio de Janeiro, São Paulo, Lima and Caracas. Over much of the continent, however, there are some very large scantily peopled areas.

We come now to the vast towns, the super-cities of today. New York, London and Tokyo have 11 millions, Paris and Buenos

Aires 8, and Chicago, Los Angeles, Shanghai and Moscow 7. Bombay, Calcutta and Peking are the next biggest, and they may soon be similar to Chicago in size. These figures are estimates. All these towns are growing, and if we include settlements round about, which may soon be swallowed up by them, the figures may well be bigger. It is difficult to explain exactly this remarkable growth. Perhaps the truest explanation is that this is the way in which people seem to wish to live. Certainly we can say that these vast cities are made possible by modern technical advances. You have studied some of the problems of urban life. Fresh water, food, power supplies, health services and good transport are but a few of the things which must be provided. They sound simple enough, but it requires a high degree of organisation to ensure that they are always available, and there would be very serious trouble should they fail.

Finally, you should notice that there is still further 'gathering together'. The chief example is in the United States. From the coast of New England, through New York and its suburbs to Philadelphia, Baltimore and Washington is a string of towns housing 40 million people, which one geographer has called Megalopolis. Smaller examples are the Ruhr, which we have mentioned, and the group of British towns round the southern Pennines. Another group is growing on the southern side of Japan, round the shores of the 'Inland Sea', and if we look into the future it is not impossible to think of a similar group in Brazil stretching from São Paulo to Rio de Janeiro.

Work to do

1. *Draw a graph to show the percentage of world population living in towns of over 5000 people (page 120). The horizontal scale could be 2 cm to represent fifty years. The vertical scale could be 2 cm to represent 10 per cent. In each column write in the figures in millions for the date concerned. Give your graph a title.*

2. i) *Trace or draw an outline map of South America. Using the world map (fig. 5.15) to help you, mark in and name towns with over 1 million inhabitants. Use a different symbol for towns with over 5 million inhabitants. Put a key and give your map a title.*
   ii) *Make a list of the towns, naming the country in which they lie, getting this information from your atlas.*

3. i) *What is meant by rural population?*
   ii) *Name two countries in South-East Asia in which the population is mainly rural.*
   iii) *In what type of settlement do most of these people live?*
   iv) *Using pictures to help you, describe the houses of these settlements.*
   v) *What crafts are carried on in these settlements? (page 158).*

4. *Use your atlas to help you answer these questions on South America.*
   i) *Name the mountains with a scanty population.*
   ii) *Name the west coast desert.*
   iii) *Name the equatorial forest area with a scanty population.*
   iv) *Other thinly peopled areas are: (a) Patagonia (southern Argentina); (b) Tierra del Fuego; (c) the Gran Chaco (north Argentina and west Paraguay); (d) Matto Grosso (south-west central Brazil). Find these in your atlas and list them.*
   v) *Draw a map of South America locating the areas named in i–iv. Give the map a title.*
   vi) *Write a paragraph explaining as far as you can why these areas are scantily peopled.*

5. *Draw a map to show the position of Cairo, using your atlas. (a) Draw the coast from Gaza to El Alamein. Add the Sinai Peninsula, and the Gulf of Suez. (b) Mark in the River Nile from the region of Asyut, putting in the delta distributaries. Name the River Nile at its southerly point in your map. (c) Put a large dot for Cairo, writing the name to the west of the dot. (d) Mark in the Suez Canal. Draw the symbol used and name the canal in the key. (e) Add small dots for Suez and Port Said. Name them in small printing. (f) Name the Mediterranean Sea. (g) Add the 30° N line of latitude. (h) Give your map a title.*

Bush airfield: Colombia. ▶
(*Shell*)

Chapter 6

PROBLEMS OF TRANSPORT

1. Roads

Man has used vehicles to carry passengers and goods for centuries. The invention of the wheel was one of the earliest steps in his development. Today the petrol engine has made road transport in many ways more important than rail, and it carries both goods and passengers. Private cars and buses for work and pleasure are well known to you. It is worth thinking about what kind of goods are carried by road. Road transport is very handy for local deliveries. On main roads you can, if you are observant, notice what other goods can be seen. Large lorries going northward out of London sometimes have bales of wool taken from the docks to Yorkshire, or loads of old tyres for remaking in the Midlands. Bulk liquids are also carried; the petrol tanker is a familiar sight. Road traffic is also speedy; vegetables and fruit are usually brought to market in this way. Lorries bringing 'primeurs' (early vegetables) from the Mediterranean south of France to Paris are often to be seen thundering along the road from Marseilles to Paris, the busy 'route nationale'.

All this traffic uses modern asphalt or concrete roads, and these are the kinds most of us who are city dwellers think of first. But if you consider other parts of only this country you will soon realise that there are many other types, and you probably know how they are marked on a map. A foot-path over the moors is hardly a road, but when it is wide enough for wheeled traffic it becomes the roughest simple road, often called a track. The first improvements to such rough tracks are sometimes made when they lead to a farm, and the farmer will fill in the ruts with some hard stone to make a firm surface. This is the beginning of the engineered road, which is built and kept in order by the local authority or the government, according to its size.

The simple country lane is a minor road, and today it is often quite wide. In hilly country areas such as Devon it can be sunk between deep banks, and passing by cars is possible only at a few places. Similarly, in remote parts of Scotland where there is little traffic, to save money only narrow roads are built, and motorists must watch the road ahead carefully and draw in to a passing place if they see another vehicle approaching. These Scottish roads are sometimes built on a soft peat surface, and can sink in places so that cars must go quite slowly over the gently switchbacking surface. Similar, but much more serious, subsidence occurs in coalfield areas, when the land sinks after the coal has been extracted. Here large signs 'Road liable to subsidence' may be seen. Apart from local roads, the large roads in this country are classified as 'B' and 'A' roads, shown in brown and red on the Ordnance Survey map. The A roads are the main system, and those numbered from 1 to 12 form our main network, radiating largely from London. Study of the map (fig. 6.1) will remind you of our main towns and regions which need connecting links. Everyone has heard of A1 as the Great North Road, or A2 as the Dover road—a route of historic importance over which for centuries travellers from the Continent have first entered Britain.

In other countries there is a somewhat similar system, but there are wider variations in type, according to the surface of the land, and even the climate. In countries which have been settled for less time than western Europe, the new agricultural lands are often crossed by what are called dirt roads. There is no hard surface—called metalling by the engineers—and the loose earth is just scraped level from time to time. At certain seasons these become impassable. In Russia, particularly on the soft earth of the steppe, the spring thaw produces much water, which flows only slowly off the level land, and roads become a sea of mud. In many tropical lands, where there is a striking difference between the wet and the dry season, roads are even labelled on the map 'all-weather roads' or 'dry season roads'. Another difficulty is experienced in the tundra lands (page 206) when the frozen soil thaws, causing the surface to subside, rather like the roads over peat in Scotland. Sometimes nature helps rather than hinders. In level dry places, such as parts of the Sahara desert, there are vast level hard or

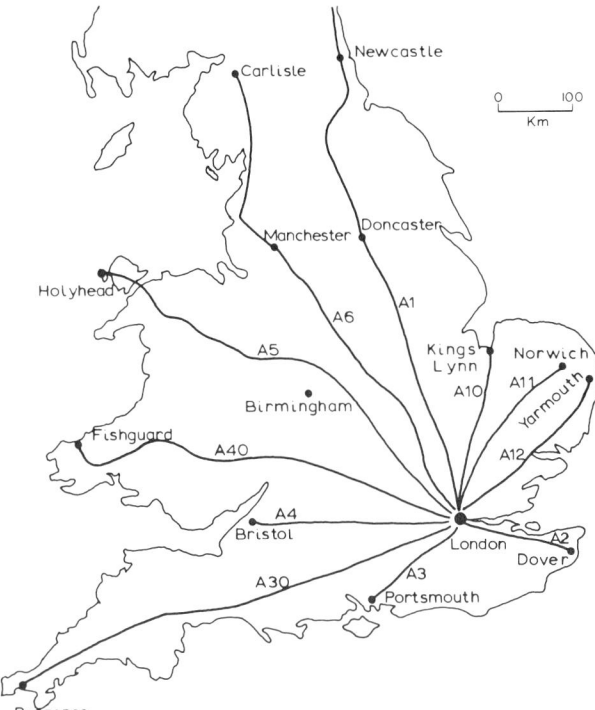

Fig. 6.1. England and Wales: A roads.

route. The main obstacles are hills, and water in the form of rivers, lakes or arms of the sea. Study the picture, and you will see how the engineers built zig-zags up the steep mountainside. Such roads involve a long, slow uphill haul, and tunnels, such as the St. Gotthard, are now being built instead. Wherever possible, roads take the level course along a river valley, and nearly all main river valleys have a main road on one or both sides of their flood-plain. When a great river valley is thus followed by roads and railways we call it a routeway. A strongly indented coastline, such as the

Mountain road: French Alps.
Explain clearly why the road is this shape.
(*J. Allan Cash*)

stony areas, and roads need not be built. During the Second World War vehicles were able to drive freely all over such areas, the way being marked by a pile of stones, or even old petrol tins. Among sand-dunes, of course, the situation was more difficult.

We have so far considered only the surface of the ground. Clearly road builders must consider the shape of the land on the

River ferry: Nigeria.
(*United Africa Company*)

Oakland Bay Bridge: San Francisco.
On a map of San Francisco Bay, mark in the two main bridges. Add a scale. What are the objects on the coast in the foreground?
(*U.S. Information Service*)

drowned glaciated coasts of Scotland and Norway, causes another problem, as roads are forced to wind inland, for tens or even hundreds of km, to cross the inlet. Ferries are often used. Where the traffic is very busy, large road bridges are built, such as the Forth and Severn bridges in this country, or the Golden Gate bridge in California.

We have not so far considered why roads should be built. They are, of course, to get people and goods from place to place, and roughly speaking, the more people there are, the more roads. After all, you can think of a town, where there are most people, as being made up of houses, shops and factories with roads in between. This is the highest density of roadways we could have. In sparsely populated areas there are few roads, as there is little need for them. Particular kinds of country have their own road pattern,

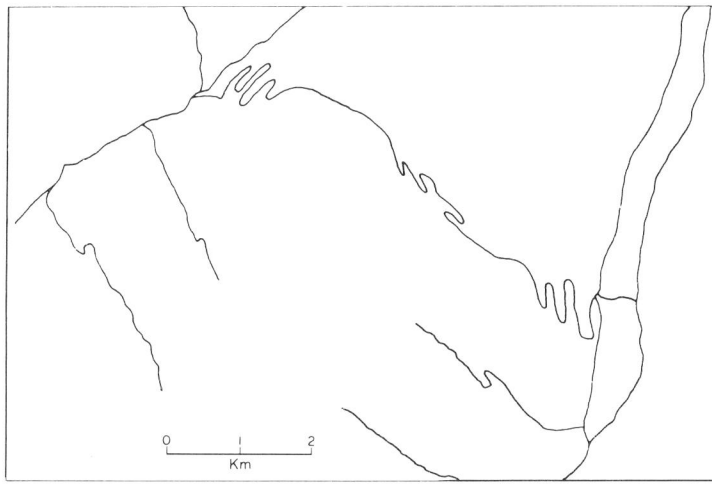

Fig. 6.2. Road pattern.

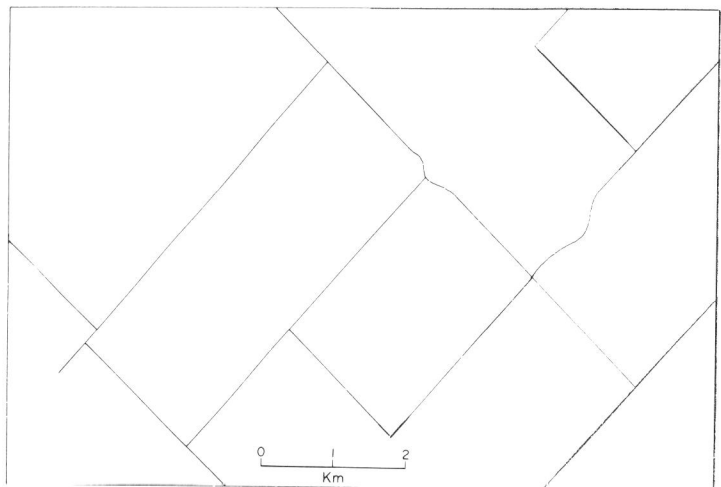

▲ Fig. 6.3. Road pattern.   Fig. 6.4. Road pattern. ▼

shown in figs. 6.2, 6.3 and 6.4. You should be able to find for yourself which is which. One is of a mountainous land, with the main roads running along the valley, or winding over the passes. Another is the newly developed flat prairie land of Alberta, and the other is in the county of Norfolk. Here farms and villages are fairly evenly scattered over the rolling countryside.

There is another kind of road pattern to be found. If you look again at fig. 6.1 you will see that the roads radiate mainly from London, the capital and largest city. Such a pattern can be seen in almost every road system, from the smallest to the largest. A view from the air of any agricultural village shows a spider's web of farm tracks, leading from the country to the village. Similarly, bigger roads focus on the nearest large town. Nowadays, as road transport has become very important, even bigger roads, or motorways, are being constructed (fig. 6.5). It was not by chance

Flyover and clover leaf: New York State.
Draw a diagram to show how this scheme works. Remember traffic
keeps to the right in U.S.A.
(*U.S. Information Service*)

that the first motorway was built to join London with the busy industrial Midlands, and rapidly extended to reach industrial Lancashire. The present pattern is of two main north–south routes, with three east–west routes linking them.

Other modern countries have built a motorway network. In Germany before the war the autobahns were built to form a network across the whole country, to move troops rapidly in preparation for war. In Italy a large motorway from the north to the south was built partly to help to develop the poor and isolated southern part of the peninsula. In America, with more motor cars than any other country, there is also an elaborate pattern of modern roads. Large and fast transcontinental buses provide the cheapest form of travel. Fig. 6.6 shows the route followed by Greyhound buses on one such transcontinental journey. Their timetable is given opposite the map. There is also an intercontinental route under construction, the Pan American Highway, which will link the United States with Central and South America.

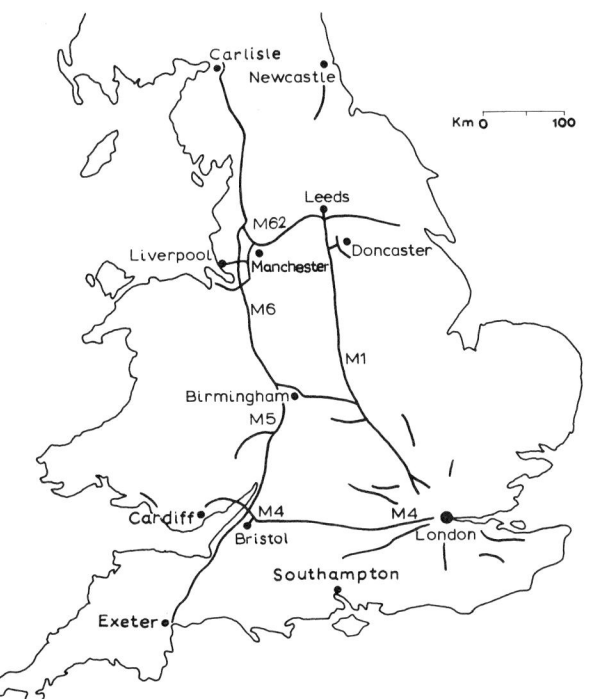

Fig. 6.5. England and Wales: Motorways open 1976.

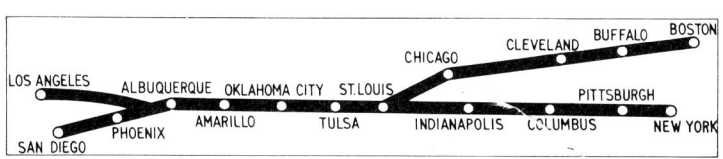

Fig. 6.6. A Greyhound bus route.

Los Angeles–St. Louis–New York

7 Through Schedules daily

		Bus 546	Bus 544	Bus 542
Los Angeles, Cal. (PST)	Lv	2.15 a.m.	2.30 p.m.	8.15 p.m.
Albuquerque, N.M. (MST)	Ar	10.05 p.m.	8.50 a.m.	2.15 p.m.
Albuquerque, N.M.	Lv	11.00 p.m.	9.45 a.m.	3.15 p.m.
Amarillo, Tex. (CST)	Ar	6.00 a.m.	4.55 p.m.	10.30 p.m.
Amarillo, Tex.	Lv	7.00 a.m.	6.00 p.m.	10.55 p.m.
Oklahoma City	Ar	1.25 p.m.	11.40 p.m.	4.25 a.m.
Oklahoma City	Lv	1.45 p.m.	12.01 a.m.	4.45 a.m.
Tulsa, Okla.	Ar		2.00 a.m.	6.45 a.m.
Tulsa, Okla.	Lv		2.15 a.m.	7.45 a.m.
Springfield, Mo.	Ar	7.40 p.m.	5.38 a.m.	11.22 a.m.
Springfield, Mo.	Lv	7.55 p.m.	6.20 a.m.	11.35 a.m.
St. Louis, Mo.	Ar	1.35 a.m.	12.10 p.m.	6.30 p.m.
St. Louis, Mo.	Lv	2.30 a.m.	1.00 p.m.	7.15 p.m.
Indianapolis, Ind. (EST)	Ar	7.30 a.m.	6.20 p.m.	12.40 a.m.
Indianapolis, Ind.	Lv	8.10 a.m.	7.00 p.m.	1.10 a.m.
Columbus, O.	Ar	1.00 p.m.	11.00 p.m.	5.10 a.m.
Columbus, O.	Lv	1.45 p.m.	11.50 p.m.	5.50 a.m.
Pittsburgh, Pa.	Ar	8.15 p.m.		11.20 a.m.
Pittsburgh, Pa.	Lv	9.00 p.m.		12.01 p.m.
Philadelphia, Pa.	Ar	3.00 a.m.	11.15 a.m.	
Philadelphia, Pa.	Lv	3.15 a.m.	11.30 a.m.	
New York, N.Y.	Ar	5.15 a.m.	1.30 p.m.	7.35 p.m.

Through Sceni-Cruiser Service Los Angeles–New York

Notes:
PST = Pacific Standard Time
MST = Mountain Standard Time
CST = Central Standard Time
EST = Eastern Standard Time

Lv = Leave
Ar = Arrive

Work to do

1. *Make a list of all the kinds of road you know, showing, where possible, the Ordnance Survey map symbols for them.*

2. *Write notes on all the difficulties which road makers find when building roads, under the headings: (a) surface; (b) relief; (c) weather and climate; (d) needs of traffic.*

3. *Using fig. 6.6 and the timetable, answer the questions:*

   i) *Locate all the towns shown in fig. 6.6, using a political map. Now write down what is meant by Cal, N.M., Tex, Okla, Mo, Ind, O, Pa and NY in the timetable.*
   ii) *At what times do the three buses leave Los Angeles?*
   iii) *How long does it take bus 546 to reach Albuquerque from Los Angeles?*
   iv) *How long does bus 544 stay in: (a) Amarillo; (b) St. Louis; (c) Philadelphia?*
   v) *Copy the timetable for bus 542. Assume that this bus left Los Angeles on a Monday. Go through the journey adding the correct day of the week for each time of arrival or departure.*
   vi) *How many days does the journey from Los Angeles to New York take?*
   vii) *Calculate the length of the journey in kilometres.*

4. *Trace a large map of some part or all of the United States, to show the towns named in the timetable.*

5. *'We are travelling through parts of Los Angeles which look more like feudal estates than suburbs, travelling through whiffs of scent from flowering gardens and parks. Sunset Boulevard goes on and on, past villages, factories, parking lots, vacant lots, past supermarkets and movie-palaces. Above all there are car-dumps—of coupés, crumpled and concertinaed like waste paper, dumps of cars which have been hydraulically flattened . . . Then there are the sinister dumps of bath-tubs, perambulators, cookers and refrigerators which would be considered modern in England. Beside many of these dumps garages and gasolene-stations advertise themselves with necklaces of bright coloured lights and cats' cradles of coloured pennants. Sunset Boulevard is as uninhibited by its ugliness as by its beauties.'*

   i) *What is Sunset Boulevard and where is it?*
   ii) *Give the English words for: (a) movie-palace; (b) gasolene station; (c) vacant lots.*
   iii) *List the items which contribute to (a) the beauty, (b) the ugliness of Sunset Boulevard.*
   iv) *Use page 5 to find out goods made in the factories. Name them.*
   v) *Try to write a similar description of a main street in your neighbourhood.*

2. Railways

Roads have been used by man for thousands of years. Railways have had a much shorter history, and today their importance is beginning to decline. They were first developed in Great Britain, during the period called the Industrial Revolution, when a great wave of mechanical inventions and the development of factory production earned us the title of 'the workshop of the world'. Strangely enough, the railways came before the engines. The first tracks were of wood, and were used in the Newcastle area to enable coal carts from the inland coal mines to run downhill to the sea. When we learnt to smelt iron cheaply, some two hundred years ago, the wooden 'ways' were soon replaced by iron rails. Other countries still call their railways 'iron ways', for example, France (chemin de fer) and Germany (Eisenbahn), though of course today they are made of the much harder steel. Steam engines were first built at the beginning of the nineteenth century, and Stephenson's Rocket, which first ran in 1829, is one of the best known of these early locomotives.

There was no plan for the building of the early railways. Small companies built and operated short lines wherever they thought it would be possible to make them pay. These lines were often to carry coal to the ports, for example, the Stockton to Darlington line, 1821, or to join two nearby towns, for example Liverpool to Manchester, 1833. It was not until the middle of the century that these lines were connected sufficiently to give continuous journeys between London and the main towns. Since then there has been steady amalgamation (joining up) of the small companies, until in 1921 there were only four, the Southern, Great Western, London Midland and Scottish, and London and North-Eastern. These groups were nationalised in 1947 to become British Railways, and our present regional divisions are largely based on the four systems taken over.

The cheap and rapid carriage of people and goods made Britain's industry prosper, but the early railway builders ran into many difficulties, whose effects remain to this day. Because they were new, and even considered to be dangerous, many people objected. As a result, building the lines was more expensive than in other countries later. Costly legal actions had to be fought, and landowners charged high prices for their land. Indeed, sometimes railways had to take longer and more difficult routes because influential people were able to stop construction near their houses or through their estates. Towns were already built, of course, and the railways were seldom built across them. In almost every town in England and Europe you will find that the railway station is some distance from the original busy centre of the town. This point is illustrated by fig. 6.7, which shows the ring of main line termini round the central part of London.

We have seen that railways are only built where there is a

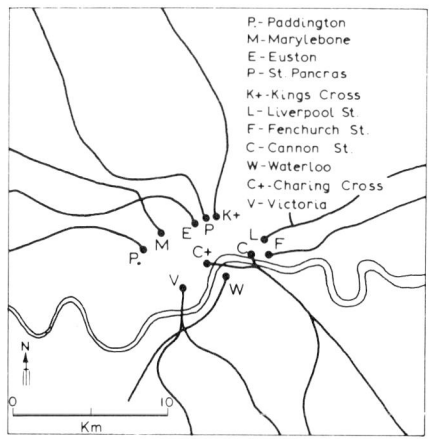

Fig. 6.7. London railway termini.

131

demand for their services, that is, to transport people or goods from one place to another. Once the two places have been decided upon, the surveyor has to plan the best route, and his problem is mainly concerned with the nature of the ground. The ideal situation occurs when this is quite flat and the line can therefore be straight and level. The longest straight railway line in the world is across the Nullarbor Plain, over 500 km, on the railway between Perth and Adelaide in Australia. More often there are hills in the way, and the route is planned to keep the gradients as gentle as possible. Steep slopes slow down heavy trains, or make extra locomotives necessary. The easiest line through hills is usually found by following a river valley up one side and down the other. If the crest is steep or high a tunnel is sometimes built cutting beneath the watershed at the top. If a long arm of the sea is in the way bridges must be built, e.g. across the Firths of Tay and Forth, or tunnels, e.g. under the Severn estuary.

In addition to following a valley, and when there are hills in the way but no valley, the track is kept as level as possible by making cuttings through hilly parts, and embankments across the lowlands. Fig. 6.8 shows an example of this. The nature of the ground affects the railway builder in another way. Some rocks are fairly easy to cut into, such as chalk. Very hard ones require much blasting and make construction more difficult. On the other hand, if the ground is soft, such as clay, it may be easy to dig out, but the sides of the cuttings would weather easily, or even slip down after heavy rain. Such cuttings have to be made with a gentle slope, and stabilised with a cover of grass and bushes. You can think of other kinds of ground which present problems to the railway engineer: loose, sandy areas, soft swampy land, lands liable to be swept by falls of rock or avalanches and many others.

By way of example, let us consider the way in which railways have crossed the Alps. Your atlas will show you what a great barrier these form. From the Mediterranean coast at the French–Italian frontier they swing in a great curve northwards and then

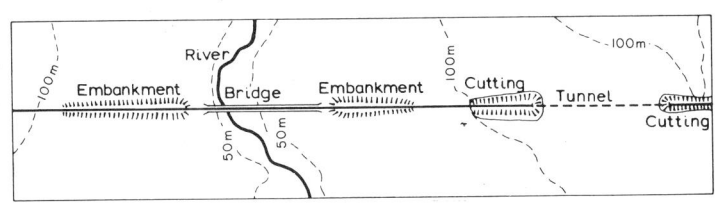

Fig. 6.8. Cuttings and embankments.

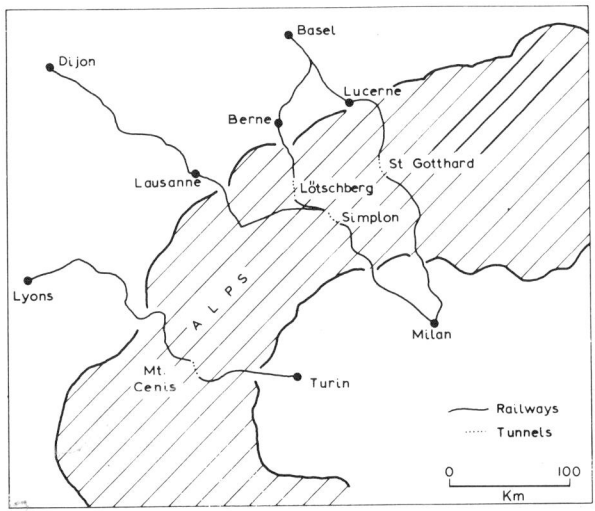

Fig. 6.9. Trans-alpine railways.

132

eastwards through France, Italy, Switzerland and Austria. In the higher western parts there are only three main railways which cross them, the Mt. Cenis route, the Simplon–Lötschberg and the St. Gotthard, and all these lines use tunnels. Farther eastwards is the Brenner route, between Austria and Italy, which is low and level enough for the railway not to have to tunnel. Study these routes in fig. 6.9, and note the main towns they join, and the river valleys they use.

The St. Gotthard route, between Switzerland and Italy, is one of the best known (fig. 6.10). When it was opened in 1882 it was one of the most advanced pieces of engineering of its day. The Alps here are at their narrowest, but they are also very high. The engineers used at first the flat land of the floor of the valley of the River Reuss. As the valley narrowed, they cut a steadily rising track along the hillside, even roofing it over in places to protect the line from falling stones and snow avalanches. Finally, just before Göschenen the railway enters the hillside in a spiral tunnel, as there is no room to build zig-zags in the narrow valley. From Göschenen to Airolo the main tunnel runs quite straight for 14 km at a height of nearly 1200 metres, far below the actual summit of the main mountain ridge. At Airolo it follows the valley of the River Ticino downwards to the Italian plain, again turning round and round in spiral tunnels where the valley is very narrow. It is now the chief railway link between Northern Europe and Italy. Some 120 trains a day use this route, carrying millions of passengers and millions of tons of goods every year. Motorists can put their cars on special wagons to go through the tunnel, and this is very useful in winter when the Alpine passes are blocked by snow.

Many boys are familiar with a railway landscape from their train-spotting days. Railways have had a considerable effect on our landscape, sometimes offering interesting scenes, but more often making the country less beautiful. You already know some of the features: cuttings, embankments, level-crossings, bridges, signal gantries and stations. One less familiar is the marshalling-

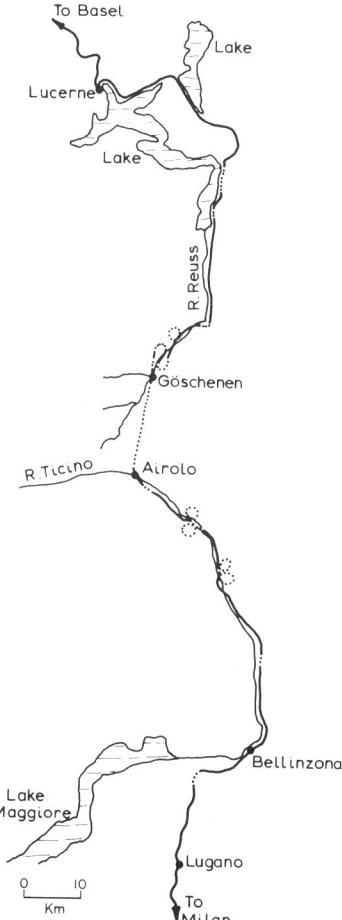

Fig. 6.10. The St. Gotthard route.

133

yard. At intervals in any railway system there must be sets of sidings, where trains and their wagons must be assembled, or rearranged according to their destination. One of the biggest is at Hamm, just north-east of the Ruhr coalfield, where train-loads of coal and other goods from the nearby industrial area are put together before their main journey. Just outside every large railway station is also such a yard, usually with engine-sheds. These are often not noticed because they are surrounded by a high fence or by other buildings. These yards take up a surprisingly large amount of land, which we can ill afford in our crowded island. Railways also need workshops, and sometimes whole towns have been built to house the work-people. Crewe and Swindon are the best known examples in England.

One part of the railway scene is changing. The steam locomotive has gone out of use. Many of our trains are hauled by diesels, but where lines have been electrified, for example, London to Liverpool and Manchester, the overhead electric conductor is a feature of the landscape and speeds are high. As you know, suburban lines are often electrified. Of course, mountainous countries with plenty of hydro-electric power also have mainly electrified lines, for example, Norway, Sweden, Switzerland, North Italy and much of France.

As more and more railways are built in a country, a railway network develops. We can find different patterns of these networks in different countries. These networks vary according to the stage of development the country has reached. You can think of these networks as the product of the history and the geography. You know now that the earliest pattern in England was hardly a network at all, but just a few short lines here and there. Fig. 6.11 shows another young network. West Africa is a developing area,

with a few separate lines each running inland from the coast. Another early form was the transcontinental line. In the pioneer days of the middle of the last century five or six railway companies built lines very rapidly right across Canada and the United States, from the settled east to the new lands of the west.

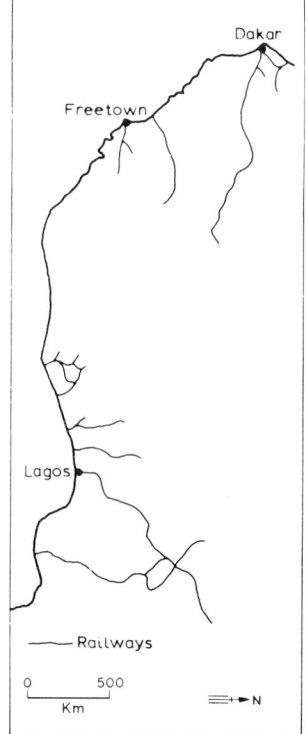

Fig. 6.11. Rail network: West Africa.

Railway yards: Swindon.
(*Aerofilms*)

135

Diesel-hauled coal train: West Virginia.
How many locomotives are pulling this train? What does this suggest?
The coal comes from West Virginia. Refer to your atlas to find its
destination.
(*U.S. Information Service*)

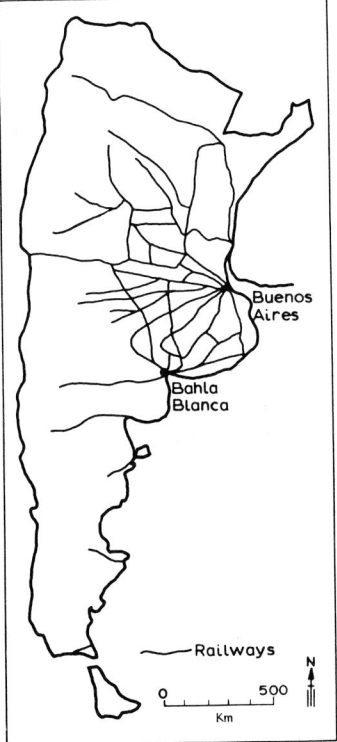

Fig. 6.12. Rail network: Argentina.

Similarly, the Russians built the Trans-Siberian railway across
Asia to open up the new lands of Siberia.

As more railways were built, the network was thickened. More
and more lines were built, particularly on flat, productive areas.
Argentina (fig. 6.12) is a good example. You can see how there is
a thick network on the level farmlands of the Pampas, and there is
a similar pattern on the wheatlands of the Prairies. Most of the
advanced countries of the world now have a fully developed rail-

way network, which is densest in the busy, populated parts, and often thinner in the mountain areas. Study your atlas map of Belgium as an example of this.

In Britain we have come to the end of the story. Our network is beginning to shrink, as road transport in certain places is cheaper and more convenient. It is easier for a factory to have its goods sent from door to door, which can be done by lorry. Only heavy bulk mineral traffic, such as ore and coal, can best go in long goods trains. Also in country districts people are best served by local buses. These are now cheaper to run than railway trains, where track and station have to be kept up for only a few passengers. As a result, British Railways are closing down many lines, mainly in country districts. This is a serious problem, for naturally the railwaymen are very concerned at losing their jobs, and at having to move house to find new ones. If you keep your eyes open you may spot quite a number of disused permanent ways in England, that is, stretches of railway track without any railway lines. There is sometimes talk that these might be made into roads, which would help solve our ever-increasing traffic problem.

Work to do

1. *Write notes on all the difficulties which engineers find when building railways, under the headings:* (a) *surface;* (b) *relief;* (c) *weather and climate;* (d) *needs of traffic.*

2. *Using the map of Brasilia* (fig. 5.12) *as a model, on an outline of Great Britain construct a similar map for London from the table below:*

London (station) to	km	London (station) to	km
(*Euston*) *Carlisle*	*481*	(*Paddington*) *Fishguard*	*420*
(*King's Cross*) *Newcastle*	*431*	(*Waterloo*) *Plymouth*	*376*
(*Liverpool St.*) *Norwich*	*185*	(*Victoria*) *Brighton*	*82*
(*Marylebone*) *Manchester*	*341*	(*Charing Cross*) *Dover*	*124*

Give your map a title. (*Omit stations in your maps.*)

3. i) *Draw a rectangle the same size as the picture of Swindon.*
   ii) *Mark in the large (dark) area of the railway lines, marshalling yards and sheds. Shade in and key.*
   iii) *The area top left is that of the latest building developments. Shade in differently and key. Include the small area top right.*
   iv) *The area to the right of the railway yards, separated off by a road, is part of the nineteenth-century town. Shade in and key.*
   v) *Give your sketch-diagram a title.*
   vi) *Write paragraphs describing each of the three areas.*

4. i) *Draw a map to show the main railway routes across the Alps in Switzerland* (fig. 6.9).
   ii) *Describe the building of the St. Gotthard route.*

5. *Write an essay on 'The problems of railways today'.*

137

3. Water transport

Man has used boats on water as long as he has used vehicles on land, perhaps longer. If you think for a moment you will easily be able to write down many different kinds of water craft. Here are a few to start you off: dug-out canoes, barges, racing yachts, cruise liners. Continue this list in your exercise book.

As with land transport, boats carry either people or goods, and sometimes a mixture of both. Tracks are not needed: wherever there is water, boats can travel. The wind gives free power, and they can be made much larger than land vehicles or aeroplanes. But water offers more resistance to movement, and except for special vehicles like hovercraft, which travel over rather than in water, boats go more slowly. They can carry cheaply but slowly enormous loads, and this is their main use today. Passenger liners across the Atlantic have disappeared, though of course there are plenty of places left where ships carry passengers. These are mainly on shorter journeys, for example, across an estuary or between islands. These boats are called ferries (fig. 6.13).

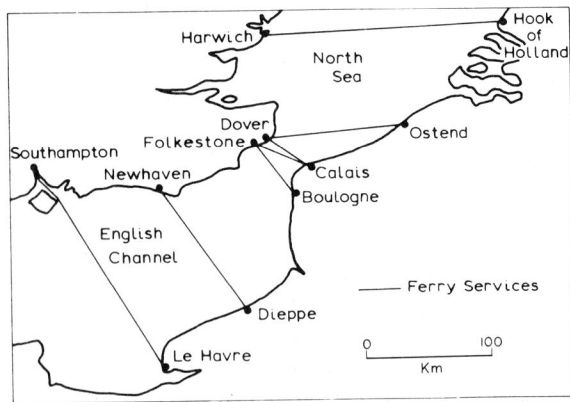

Fig. 6.13. Channel ferry routes.

Barge on the River Rhine.
What is being carried in the barge? Note that this material is carried in bulk.
(*J. Allan Cash*)

Boats use rivers, the sea and canals. Let us first consider rivers. Great rivers form natural waterways leading into the heart of continents, but they are not always important for shipping, for reasons which we shall discover. We are going to study the Yangtse Kiang (Kiang is the Chinese word for river). This is one of the busiest rivers in the world, largely because it flows through the central part of China, which is very densely populated and produces many goods to be transported. The river is over 5000 km long, and flows from the mountainous heart of Asia. In its upper course it flows between remote mountain ranges, and is of little use, though navigable in one part, the province of Sze-Chwan. This fertile province is surrounded by mountains, and where the river cuts across them it flows very rapidly through a series of gorges at Ichang, where the current sometimes reaches a speed of 20 k.p.h. For centuries junks have been hauled upstream by teams of men, using a difficult and dangerous path at the side of the gorge. Now the Chinese Government plans to build a dam across the gorge, which will create an enormous lake, and permit navigation by large ships up to Chungking. Ships will move upstream by locks. From Ichang the river flows more slowly, falling only some 30 metres on its 1600 km journey to the sea. In summer it is very deep, and ocean-going vessels of 15 000 tonnes can reach the main town of central China, Wuhan. Specially constructed river boats, which need only 2 m of water, use the river all the year round, and there are also innumerable smaller junks. Navigable tributaries also meet at Wuhan, so that this great town is a natural inland port, over 1000 km from the sea.

But you will find (page 177) that China is a monsoon land, with much of its rain in the summer. This causes great difficulties. The river level at Wuhan changes by over 12 m from summer to winter. If a ship is late leaving the town, or there is a sudden drop in the water-level, it may be stranded on a mud-bank until the next summer. In flood time also the river spreads for several km across its flood-plain, so it is difficult to find the deep-water channel, which has to be marked by buoys. It is often impossible to build wharves or landing places, because in winter they would be a long way from the water. Special landing stages must be built, connected to the shore by pontoon bridges, that is, a road mounted on barges or floats, which can settle on the mud during low water, and float in the summer flood season.

If you study your atlas you will find many other great rivers which appear to offer a waterway from the interior of continents. For various reasons, they are not as important as you might think. The mighty Amazon, the deepest of them all, can be navigated by large vessels right across South America almost to the Andes, but its basin is heavily forested and thinly peopled, so there is little traffic. The Congo is blocked by falls near its mouth. The mouth of the Murray River in Australia is blocked by a sandbar, and also tends to dry up. The Mississippi has a particularly winding course, so that the river journey is many times longer than a straight line between the towns. These curves are called meanders. They are found on most rivers and are a serious disadvantage to nearly all river transport (page 45). Before the railway age, however, the Mississippi was important in the opening up of North America, and there are many stories about its river steamboats.

The problem is sometimes not whether waterways *can* be used but whether man wants to use them. Where a great waterway offers some possibilities for easy transport man improves it if he needs to. The Great Lakes–St. Lawrence route of North America is an example. The wheat from the Prairies (page 200), iron ore from Wisconsin and Labrador, and the coal of Pennsylvania are bulky cargoes which man needs to move cheaply. In these cold lands the water freezes in winter, and this problem is partly overcome by ice-breakers. Even so, the route is closed for five months each year. There are also rapids and falls. The best known are Niagara Falls (fig. 2.19). The most recent improvement was the construction of canals round the rapids in the St. Lawrence River, which completed the St. Lawrence Seaway and enabled ocean-going vessels to sail inland as far as Duluth (figs. 6.14 and 6.15).

The most important waterway of all is the sea. Ocean-going ships can be very big indeed, though as we shall see, there are limits to their size. Ships could at one time be divided into passenger liners, which ran to a timetable, and the tramp steamers, which roamed from port to port collecting cargo as it was available. Today this is changing. Most passengers travel by air, and the liners are being used for holiday cruises. Much cargo is now carried in container ships, which saves dock labour, or in special large bulk cargo vessels. Perhaps the most important cargo in the world today is oil, and special tankers carry this and nothing else.

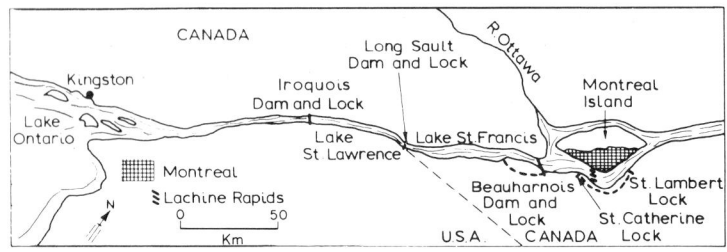

Fig. 6.14. The St. Lawrence Seaway.

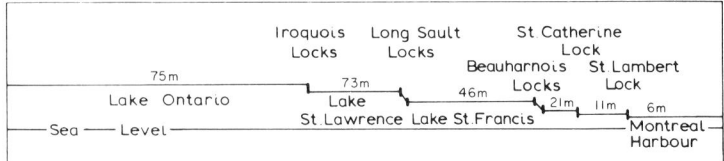

Fig. 6.15. Locks on the St. Lawrence Seaway.

particularly stormy (fig. 1.3). At sea, also, a ship must know where it is, and this is the problem of the navigator. He works out the latitude by observing the angle above the horizon of the sun or a star, and the longitude by comparing his local time with that of Greenwich, that is, longitude 0. When he is approaching land, especially at night or in fog, the radar gives him on a screen a 'picture' of the coast. Finally, he needs harbours, and these were considered in Chapter 5.

We come finally to canals. These need fairly level land, for they are very expensive to construct over hills, and small ones are now no longer used. There are some quite large canals on the level plains of northern Europe, for example, the Albert Canal in Belgium. The only really important canals nowadays are those which take ocean-going ships, and fig. 6.16 shows the position of the two canals which are still of world importance. You can see that the Suez Canal shortens the journey to the Indian Ocean from the Atlantic, and the Panama shortens that from the Atlantic to the Pacific Ocean.

The surplus oil discharged at sea when tanks are cleaned, or even after a wreck, presents the very great problem of beach pollution. You may unluckily get what looks like tar on your clothes or feet at the seaside; it is discharged crude oil. The refrigerator ship carrying meat, butter and other perishables was originally invented to deal with the problem of carrying food from the producing areas in the southern hemisphere across the warm tropics to the northern lands. Not all ships are big, nor are all harbours, so there are also small coasting vessels for local trade.

The speed of sailing ships depends on a favourable wind. Today this is not important, but captains at sea still try to avoid storms, and the international weather reports help them to do this. Listen to the shipping forecast on your radio and notice how much mention is still made of winds. Certain parts of the world are

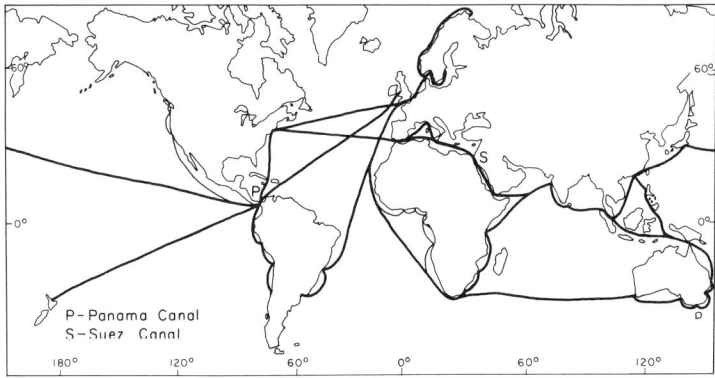

Fig. 6.16. World sea routes.

The Panama Canal (fig. 6.17) was exceedingly important to the United States; look at the distance saved on a sea journey from New York to, e.g., San Francisco. There were two main problems. The area was hot, wet and mosquito infested, and it was not until disease could be controlled that the canal could be built. The other problem was the Culebra Mountains. The Chagres River was dammed to make a lake, and ships are lifted up to this lake by the Gatun locks. They sail across this lake and then pass through the Culebra cutting, where they go down by other flights of locks to the sea again. The United States obtained the right to build the canal from the state of Panama, and takes great care that this international waterway is kept open.

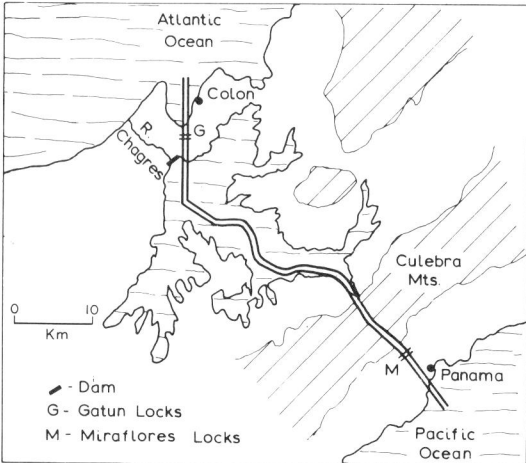

Fig. 6.17. The Panama Canal.

◀ **The Amsterdam–Rhine Canal.**
Draw a large field sketch of this picture labelling the locks, the bridges and the smaller canal which is crossing the main one.
(*Shell*)

Giant oil tanker.
This is a modern 200 000-tonne Japanese-built oil tanker. Large depots
are being built in inlets of our west coasts for the transhipment of oil
to smaller ports in western Europe.
(*Shell*)

The Suez Canal which was closed between 1967 and 1975, is now re-opened but is no longer so important. Most of the previous Suez Canal trade consisted of oil tankers from the Middle East to Europe, and the new super-tankers of 200 000 tonnes are too large to go through the canal anyway. A very large part of the world's sea trade is oil, and it makes us think of the future when we realise that tankers of 300 000 tonnes will unload at Bantry Bay or Milford Haven because they need more than the 20 metres of water which is the safe depth for shipping in the Straits of Dover.

Work to do

1. *Using your atlas, draw a map of the Yangtse Kiang.*

   i) *Rule two lines each 10 cm long, 4 cm apart. Label the bottom line 23½° N., the other 35° N.*

   ii) *Between them, locate accurately and draw in the river. Remember that although the river is 5000 km long, your map is roughly on the scale of 500 km to 1 cm.*

   iii) *Add the coast between these latitudes, showing where the river enters the sea.*

   iv) *Mark in lightly the 400 metre contour and that for 1800 metres.*

   v) *Shade in land over 1800 m with close diagonal lines; land between 400 and 1800 m with wider-spaced diagonal lines. Leave land below 400 m unshaded.*

   vi) *Mark in and name Ichang, Chungking, Wuhan and Shanghai. Give your map a title.*

   vii) *Write a paragraph to describe difficulties of transport on the Yangtse Kiang.*

2. i) *State the hindrances to transport on* (a) *the Congo,* (b) *the Mississippi rivers.*

   ii) *Why is the Murray River (Australia) not much used for transport?*

   iii) *For how long is the River Yenesei (U.S.S.R.) navigable? Why? (page 207).*

   iv) *List all the hindrances to navigation on the Great Lakes–St. Lawrence waterway.*

3. *'Most Mississippi boats are now steam or diesel screw-driven craft, and they push (not pull) enormous loads of modern barges anywhere from Pittsburgh to Texas . . . Sometimes they push a miscellaneous collection of barges, lashed together shapelessly, piled high with coal or yellow sulphur; sometimes a line of 'integrated' barges, made to fit each other, and generally containing oil. Occasionally you may see a triple-decker barge, carrying cars downstream from Detroit . . . All this traffic moves in an unceasing stream from the industrial regions southwards, and from the southern oilfields up to the Middle West. Pittsburgh, on the upper Ohio, is an important river port, and so is St. Paul, more than 3000 km from the Gulf of Mexico.'*

   i) *What phrase tells you that the Mississippi is important for transporting goods?*

   ii) *Name the four cargoes mentioned.*

   iii) *What industry uses sulphur?*

   iv) *How is Pittsburgh linked to the Mississippi?*

   v) *For what manufacture is Detroit famous?*

   vi) *Attempt to draw a small map to illustrate this passage.*

4. *On an outline map of the world:*

   i) *Mark in and name:* (a) *the Panama Canal;* (b) *the Suez Canal.*

   ii) *These sea routes use the Panama Canal. Mark in and name the ports; add the routes:* (a) *New York to San Francisco;* (b) *Wellington to Liverpool;* (c) *Callao (Lima) to Rio de Janeiro.*

   iii) *Mark in an oil route from Kuwait to Rotterdam, avoiding the Suez Canal.*

   iv) *Give your map a title.*

5. *Using the map* (fig. 1.3) *to help you, write a brief essay on 'Weather hazards at sea'.*

4. Air transport

There is little need to emphasise the importance of air transport in the modern world. Most schoolboys, if not schoolgirls, have at their fingertips details of the latest jet aircraft, or of the performance figures of helicopters and hovercraft. It was in the 1950's that more passengers crossed the Atlantic by air than by sea. This marks a turning-point in world communications. It is also cheaper to cross the Atlantic by tourist aircraft than by passenger liner. Although we find air routes in almost all parts of the world, it is possible to make certain general statements about them. The advantage in speed of aircraft over boats, compared with the advantage of aircraft over cars or trains, makes journeys over water or undeveloped desert or forest areas particularly important. There are main air routes over the great oceans, but we should remember the importance of aircraft in linking islands, for example, the West Indies or the Hebrides (fig. 6.18). We will consider the main pattern of world air routes later.

Most aircraft space is taken up with passengers, and you hear and see most about this kind of traffic. It is worth remembering that goods are also carried. Measured by the cash value of the items, Heathrow is the third port of the United Kingdom, after the seaports of London and Liverpool. At present, as air transport is expensive, we do not find heavy bulky items such as iron ore or wheat. The most obvious example of a very valuable yet small item is gold or diamonds. Nowadays these are nearly always shipped by air. Other goods carried are small, intricate pieces of equipment and machinery, such as radio parts, watches and other instruments. Women's dresses, after being bought at fashion shows in Paris or New York, are also valuable enough to be sent by air. Air transport is very rapid, and emergency supplies are moved by air; examples of these are special vaccination serums or blood supplies, or spare parts for vital machinery which has broken down. In this connection you should remember that air

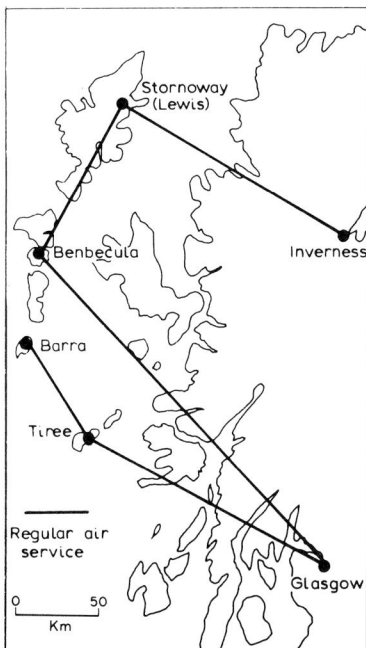

Fig. 6.18. Air links with the Hebrides.

transport, particularly of machinery and equipment, is becoming increasingly important in helping to develop remote areas (page 150). The far north of Canada, with its increasing mineral production, uses aircraft a great deal.

Not all goods moved by air do so at economic rates, that is, at prices which their senders can afford to pay and still make a profit. In 1947 when the U.S.S.R. Government closed the road and railway from western Germany to Berlin a large-scale airlift kept the isolated city supplied with all its essential requirements. Similarly, the British Government has paid large sums to move oil, which normally is not transported by air, into Zambia.

Fig. 6.19. Airways in England and Wales.

There are some technical matters which you must know before you can fully understand the geography of air transport. These technical considerations are changing rapidly and constantly. For example, at present airports need long runways, and must be very large. But if vertical take-off machines become really cheap and replace ordinary aeroplanes this factor affecting the size of an airport will change. Modern aircraft are getting larger and more expensive daily. One million pounds is now not an unusual price for an aeroplane. This means that only very large companies or even governments of countries can enter the air transport industry. You should notice also that the kind of journey affects the payload, that is, what the aircraft can carry besides the crew and fuel. On a long trans-ocean flight much more petrol must be carried, and this reduces the amount of cargo. Where short flights are possible, with refuelling halts, more can be carried, and costs are lower, though you must remember that there will be extra airport charges to be met. Countries cannot run airports for nothing. There are also vast technical requirements in connection with the control of aircraft, usually run by the Government of the country concerned. Radar stations must be manned, and information on aircraft movements fed to the control towers of the main airfields (fig. 6.19).

145

Over busy airports the air-space is becoming increasingly full up, and if necessary, planes have to wait their turn to land. They are held by the air control officials at different heights above the airport, a process known as 'stacking' (fig. 6.20). There are also many extra technical services needed on the ground at airports. The chief is the large engineering workshops to service the aircraft, but offices, petrol stores and supply vehicles, and kitchens to supply pre-packed meals, all take up much space. Besides these are the usual public waiting rooms, customs and baggage offices.

These are technical and economic factors which affect air transport. There are other matters, already familiar to you from geography lessons, which we must consider. The first is the weather, that is the conditions of the atmosphere. It is clear that weather conditions affect air travel very much, and there is a network of meteorological stations which provides international information (page 6). One obvious fact a pilot needs to know is about severe storms or headwinds. Large thunderstorms with their strong convectional currents (page 173) can cause much air sickness or even damage the aeroplane. Visibility, particularly on landing, must also be known. The amount of dust may be so high as to obscure the approach, but of course fog on landing is the greatest danger, and accurate forecasts must be made. There are various situations which produce fog (page 5). It is caused by the condensation of water vapour on minute particles of dust, so that water droplets are formed. This happens when the temperature falls and the air is still, and it is the chief problem of the forecaster to try to find this critical temperature. As we go upwards from the earth's surface, the temperature falls, and the condensation of water vapour may be directly in the form of ice. This is a serious hazard for aircraft, as a thick deposit of ice makes them heavier, or may even cause working parts to freeze up. Most planes have de-icing equipment, but still great care must be taken. You also know that at high altitudes the air is less dense. It therefore gives less support to a plane's wings. Aircraft taking off from an airport at a high altitude can therefore carry less load than when leaving from sea-

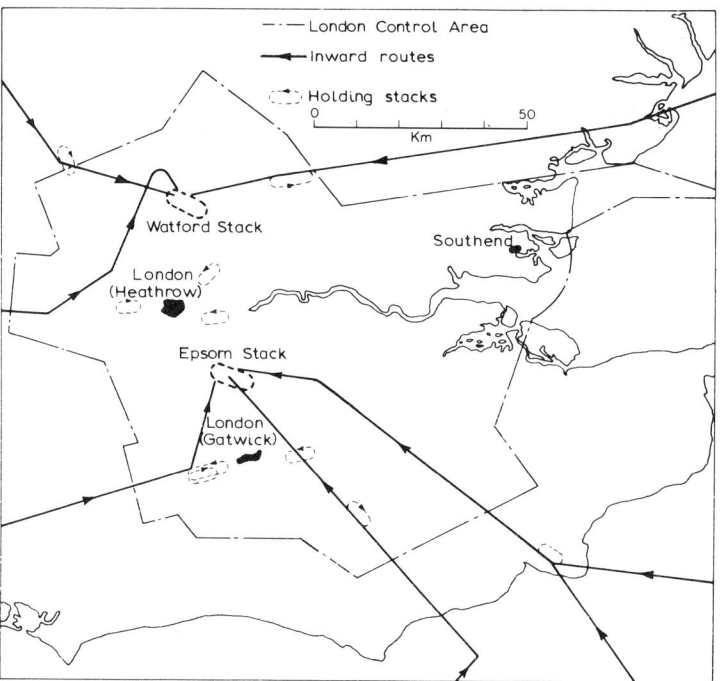

Fig. 6.20. London airport: stacks and control area.

level. This less dense air at high altitudes causes discomfort, and aircraft hulls are therefore made airtight to keep the normal ground-level pressure for the passengers.

You probably think of winds as being the most important weather feature. Certainly prevailing winds are important. The early Atlantic crossings were made from west to east, helped by the fact that in general winds across the North Atlantic are from the south-west. For many years a west-bound crossing was more difficult. But you know that nowadays planes fly at great altitudes, often at 10 000 m and more. This means that they can fly

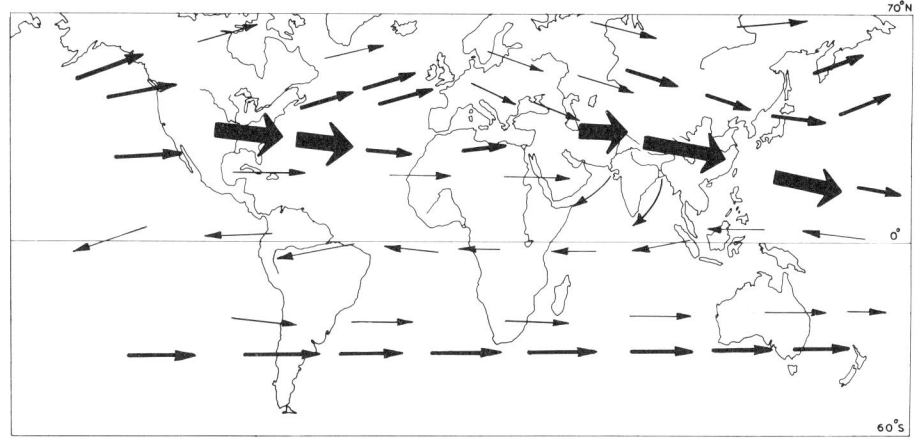

Fig. 6.21. World air circulation at 10 000 metres.

above much of the surface bad weather, and more notice is now being taken of air currents in the upper atmosphere. Fig. 6.21 shows the general circulations of the air at 10 000 m. The westerly current is in places called the jet stream. This is one of the newest discoveries in meteorology, and when it is very strong, aircraft, of course, avoid it.

The great advantage of aircraft is that they do not need any tracks on the ground, and they are in general independent of the surface of the land. This does, however, affect air routes to a certain extent. Aircraft can nowadays fly over the highest mountains quite easily, but they are still a hazard. Crashes still occur when an error of altitude is made, and the plane flies into instead of over a mountain. There is usually more cloud, and rather 'bumpy' weather over mountain ranges, and planes avoid them if possible. Landing grounds are not easily found. Innsbruck airport

in Austria (fig. 6.22) shows a typical problem. Planes must fly very carefully along the valley during ascent or descent, in one of two directions. Fortunately the valley funnels the wind so that it is also usually in one of these directions. You know that planes take-off or land into the wind. In the past small planes flew with, if possible, an emergency landing ground always within reach. For

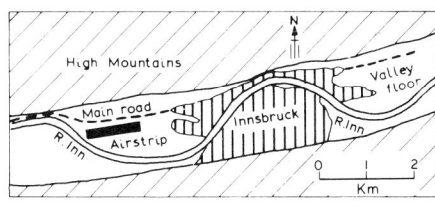

Fig. 6.22. Innsbruck airport.

this reason planes fitted to land on water could explore the Canadian north, with its many lakes, and early African services used flying boats, which could land on great lakes or rivers. The large planes of today need prepared airfields, and emergency landings would be serious indeed. Nevertheless, the great empty lands, the deserts, the polar regions and the great forests are still avoided when possible, because survival and rescue from such places would still be difficult, even if a safe landing was made. Small aircraft can still develop air traffic best over certain types of land. Simple airstrips can be most easily constructed on flat grasslands and deserts, and only with difficulty in forests. This is shown by the development of the military helicopter, which can conduct operations from the smallest jungle clearings in places like Borneo. The wealthy Texan cattle or oil man is often portrayed with his private plane on a runway near his ranch.

The surface of the land is most important in the construction of great airports. Before considering this let us study London's great airport at Heathrow. It covers over 10 square km, and is situated about 20 km west of central London. There are at present five main runways, arranged so that planes can take off whatever the direction of the wind. As the ground is very flat, special reservoirs exist in case the rainfall is more than the natural run-off. The main buildings are in the centre, connected by tunnels to the main roads. The control tower, with connected navigational and wireless equipment, is the hub of the whole airport, rather like the bridge of a ship. There are three main passenger buildings, with restaurants, waiting rooms, and customs offices; one is for British, one for European and one for Intercontinent traffic. Round the perimeter are the 'working' buildings: offices and engineering sheds for the various companies, petrol firms and their storage space, and a fire station. A new special area for cargo is being built.

You can see from the details given of Heathrow what is required to develop a modern airport. Clearly it must offer a wide level space, where long runways can be built. The land must be

Heathrow airport.
Identify on this picture the features of an airport mentioned in the text.
(*Aerofilms*)

firm and well drained, for example, of gravel, to support the heavy planes. Runways are expensive to build on soft marshy ground. The approaches should also be clear of hills or tall buildings. Large towns can create much smoke from their chimneys, so it is useful to be up-wind of them, that is, on the side which does not normally have smoke blown over it. Fog most often collects over low-lying ground, so a slightly raised site is best. We have also

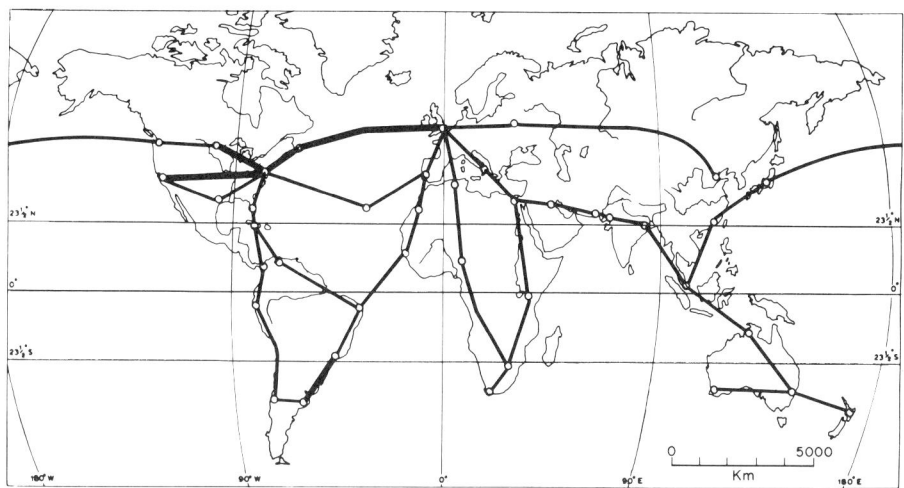

Fig. 6.23. World air routes.

seen how many large buildings are needed, so there must be space for them. Finally, the most important thing is a good connection to the centre of the town. This is particularly vital for short journeys. It is little use travelling from, say, Paris to London in less than an hour if another hour is spent getting from the airport to the city centre. Partly for this reason, the M4 motorway was opened in London. Another difficulty today is noise. Jets when landing are very noisy, and people on the outskirts of a town, which is often the best place for the airport, try to get new airports built farther away in open country.

We can now consider the actual air routes. You have seen that there is a dense road network where there are many people, and to a certain extent this is true of airways. In countries with many large towns, and a prosperous population, such as in Europe or the United States, there is a dense network of air services. We have also seen that there are often small local airlines to islands, or to some particularly remote places. The main world pattern of air routes is shown in fig. 6.23. These are the main inter-continental routes, and London, which is roughly central to the land hemisphere, is well situated to be the busiest air centre of the world. The main exception to this is for trans-Pacific flights. Notice on a globe how a flight over the North Pole is the shortest way to the west coast of Canada, and the route via Iceland and Greenland the shortest way to San Francisco. In these cases, however, the extra difficulties of flights across the polar lands must be considered.

From the map you might think that aircraft can fly freely all over the world, apart from the problems we have considered.

There is one more. Countries have control of the air above them, and not all have agreed to allow foreign aircraft to fly freely over their land. Some still refuse to allow foreign aircraft to land for passengers or trade, unless they have similar rights in the other country. These are problems of international agreement which are by no means settled.

Work to do

1. *Choose three items of cargo and explain why each can be sent by air.*

2. *Write a paragraph explaining why it is cheaper to cross the Atlantic by tourist aircraft than by passenger liner.*

3. *'The primary need was for the construction of a railway between the iron-ore field and Seven Islands, for no navigation was possible on the rivers and no other form of surface link existed. Aircraft supplied material for the construction of the railway and of four base camps. The variety of goods that had to be transported may well be imagined when one realises that from 1948 until 1954, when the railway was completed,* all men, provisions and equipment had to go by air. During the short summer season when air traffic was possible, a continuous day and night shuttle service was maintained.'

   i) *Read this passage, then read about the same area on page 25. Where is the area?*
   ii) *Why were aeroplanes essential here?*
   iii) *Why was air traffic possible only in the short summer season?*
   iv) *What men were taken by air for work on the railway? Name three essential occupations.*
   v) *What is meant by provisions? Make a list of those which were essential.*
   vi) *Equipment for railway lines does not usually go by air. Why not?*
   vii) *What settlement eventually grew up at the iron-ore mines?*

4. *Make a list of the needs of a good site for an airport.*

5. i) *Give three reasons why airports are not popular with people who have to live near them.*
   ii) *Give three reasons why airports are built near large towns.*

India's millions: Calcutta. ▶
(*Camera Press*)

Chapter 7
PROBLEMS OF POPULATION

1. World population

Information about the population of countries is collected by means of a census. A census is really a counting-up of people. In our country and many others in the world householders, hotels, boarding houses, hospitals, nursing homes and so on are sent by post a census form, a piece of paper with various questions to which the householder or person in charge has to give the answer on a given date. He has to write down the names of all people sleeping in his house or building on that particular night, and give various other items of information about them. He then posts the form back to the government department concerned. In countries where the majority of the people cannot read or write government officials visit all the houses themselves to collect the necessary information. This takes a long time. It also takes a long time to put together all the information gathered, whether by post or by visiting, and censuses are commonly taken only every fifth year, though some countries do not have them so often. We cannot say that the information gained from censuses is completely accurate, for not all people fill in their forms properly, while in some countries it is almost impossible to collect information about, say, nomadic tribes or primitive peoples living in remote jungles. Nevertheless, census returns give at least a good idea of the numbers of people living in the world. In 1964 this was estimated to be 3250 millions.

These 3250 million people are not spread evenly over the land surface of the world. 50 per cent live on the Asian mainland from northern China to the western limits of the Indian subcontinent, including the fringing islands. A second great belt of people, 20 per cent of the world's total, lives in Europe, including U.S.S.R. as far east as the Ural Mountains. A third belt, roughly 5 per cent, lives in eastern North America, particularly between the Great Lakes and the Atlantic Ocean. You can see from the map (fig. 7.1) that this three-quarters of the world's population lives in the northern hemisphere mainly between latitudes 60° N. and 20° N.

These are the areas with temperate climates, or those at the northern limits of tropical climates. There is much land north of 60° N. in Asia and North America, but few people live there, for reasons which we shall discover. The lands where many people are concentrated together are called lands with a high density of population. The remaining quarter of the earth's population lives either in smaller areas of high density, such as the western United States, south-eastern Australia, the la Plata estuary lands of South America and the north and south coastlands of Africa, or spread thinly over the rest of the southern continents in what are called low-density populations. Notice again that the smaller high-density populations lie in the temperate lands.

One reason for the low density or sparseness of population in large areas of the world is severe cold, man's worst enemy. In Canada, for example, the northern limit of the populated areas in places coincides with the January isotherm of −20° C. Modern scientific inventions have not removed the difficulties of living in very cold lands, they have merely helped man to overcome them to some extent. Thus modern insulated buildings, heating systems for houses and pavements, and ultra-violet lamps which provide artificial sun-rays have enabled the Russians to create new towns in the Soviet Arctic—Kirovsk, Igarka and Vorkuta (page 207). The Canadians and Americans have also built new towns, some completely underground, in their Arctic lands. Nevertheless, despite the introduction of special strains of grain and vegetables which will grow in the short, cool Arctic summer, special vitaminised food has to be imported for all the townspeople, and it is unlikely that the Arctic lands will ever be densely peopled.

Mountains, too, may be areas of low density or scanty population. The height at which humans can live varies a great deal. In the Alps very few people live permanently at a height of more than 2000 metres. On the other hand, the Andean Indians of South America are used to living at altitudes ranging from 3000 to 5000 metres. If we went straight from Britain to the heights of the Andes we should probably suffer badly from mountain sickness.

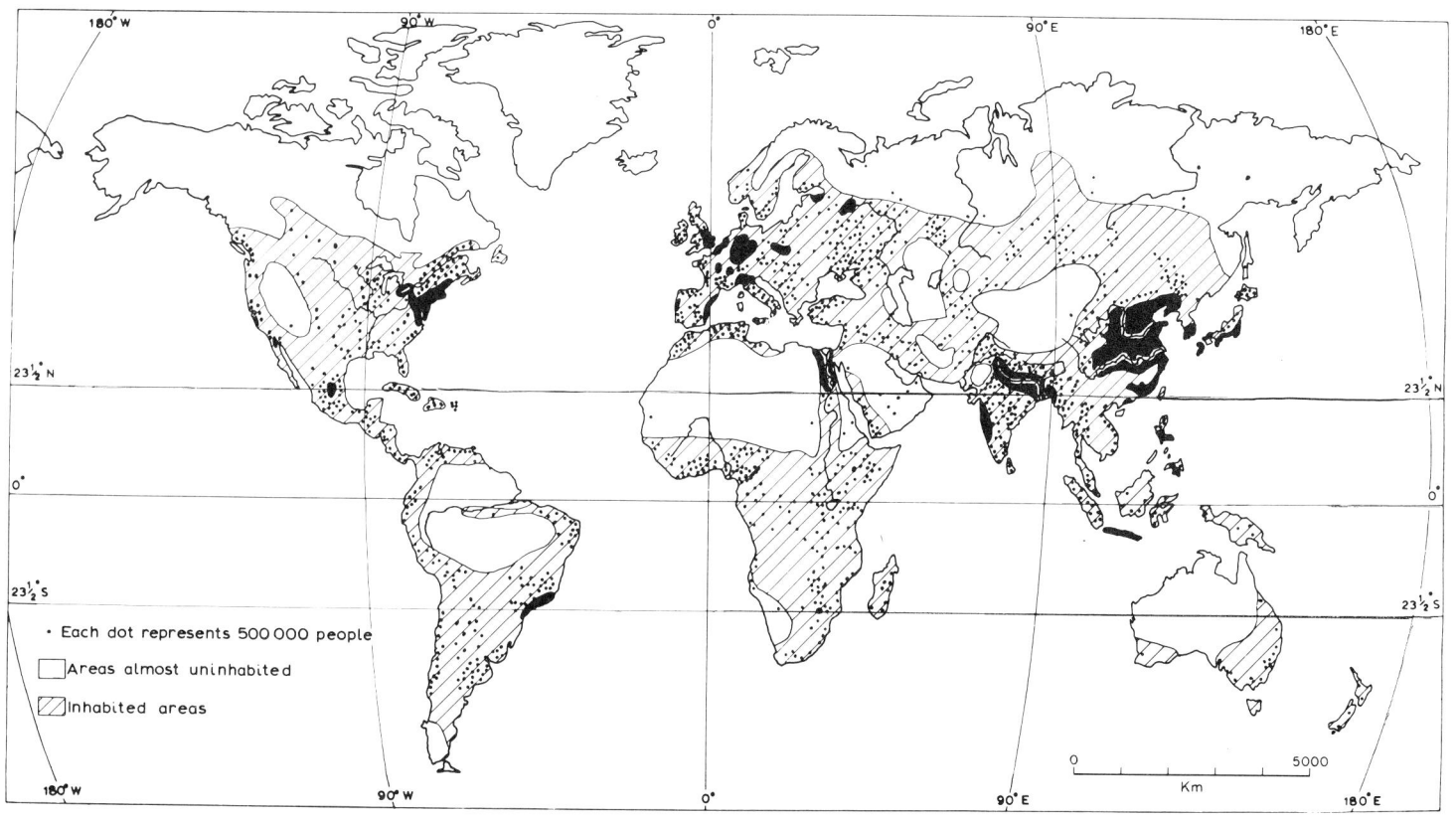

Legend on map:

• Each dot represents 500 000 people

Areas almost uninhabited

Inhabited areas

180°W   90°W   0°   90°E   180°E

23½°N   0°   23½°S

0   5000

Km

Fig. 7.1. World distribution of population.

This is a malady caused by breathing the thin air of mountain heights; a feeling of nausea which attacks many people at unaccustomed altitudes. To overcome the possibility of mountain sickness, mountain climbers train so that they are physically very fit; even so, they may carry oxygen cylinders to provide them with extra oxygen on the highest peaks. Our Olympic athletes went to Mexico some weeks before the Olympic Games took place in 1968 so that they could get used to the altitude of the Mexican plateau before competing. If mountains lie in great ranges, with sharp ridges, steep slopes, deep gorges and bare rock outcrops with treacherous scree they are difficult for man to settle in whatever their altitude. High mountain peaks or plateaux with permanent snow and ice present areas of cold climate where man cannot live. If, however, high mountains lie in tropical areas or hot deserts they may provide welcome areas of lower temperatures where people may live more easily. In Equador 85 per cent of the population live in the Andean mountain ranges; in Bolivia the figure is 75 per cent, and La Paz, at 3647 metres, is the world's highest capital city.

Deserts are lands of scanty population owing to lack of water. We discuss deserts in Chapter 8, so it is only necessary to remind ourselves that many desert dwellers are nomads, though dense populations may dwell in the small areas of oases. Man's problem in the desert is the provision of water. Here again modern devices, such as underground canals in the Israel Negev and very deep boreholes in the Sahara, are helping man. The settlement of deserts will always be controlled by the amount of water available.

Areas of equatorial forest, always hot and wet, are equally unfavourable to settlement. Here man can only just survive, in a constant struggle against excess heat, excess rainfall and disease. Fortunately not all lands on or near the Equator are densely forested. In South America the equatorial Andean areas of Colombia, Equador and Peru have more temperate climates; so do the highlands of Uganda, Kenya and Tanzania in Africa. In

Standard of living: nomads.
These are Bedouin Arabs, who move about the desert with their flocks (page 189). Why can they have only few possessions? What do you think these are?
(*J. Allan Cash*)

these areas more people are able to live. The equatorial islands of Indonesia, particularly Java, also support many people (page 175).

We have considered those parts of the world where man does not live in large numbers. Perhaps more important are the areas of high-density population mentioned on page 152. South-East Asia has the greatest rural population densities in the world, from 350 to 1000 per square km. These people depend largely on successful rice cultivation. Rice is one of the world's most useful plants (page 186). It flourishes where there is plenty of moisture during a hot growing season, but needs a dry harvest period. This means that tropical monsoon climates suit it well. It will grow in temperate climates if the summer is hot enough. It will grow in a variety of soils, producing more than twice as much grain per hectare as wheat. It has high food value, especially when eaten in the husk, for this outside cover of the grain contains fat, minerals and vitamins. Rice is therefore the cereal which normally prevents starvation among the huge populations crowded on the lowlands of South-East Asia.

The high densities of population in Europe depend less on the production of grain than on industry. The areas of densest population, apart from those of capital cities, are generally to be found on or near coalfields. This is because coal was the source of power used in the industrial developments of the nineteenth century. Nowadays new sources of power, such as hydro-electricity, oil and natural gas, enable industries to be more scattered or dispersed, but they are still centres of dense population. In western Europe the greatest densities lie within a circle of 800 km radius centred on Brussels. You can draw a map to show this circle, with Glasgow, Copenhagen, Vienna, Marseilles and Bordeaux lying on its circumference. In the U.S.S.R. the greatest number of people live in the Moscow industrial area and that of the Donbas coalfield. The country with the densest population compared with its area is the Netherlands, which averages 342 people per square km (fig. 7.2). It is both an agricultural and industrial country, but 863 000 people live in Amsterdam and 728 000 in the port of Rotterdam. Iceland, the least densely populated, has only 2 persons per square km (fig. 7.3).

The densely peopled lands of eastern North America contain many great towns. Find in your atlas Quebec, Montreal, Pittsburgh, New York, Philadelphia, Baltimore and Washington. You will notice that all save Pittsburgh and Washington are ports. There are also hundreds of smaller towns like Ottawa, Toronto, Buffalo and Birmingham. All have industries which together

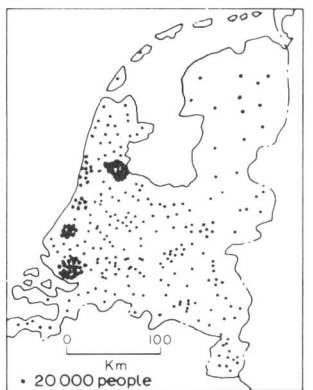

Fig. 7.2. Population density: The Netherlands.

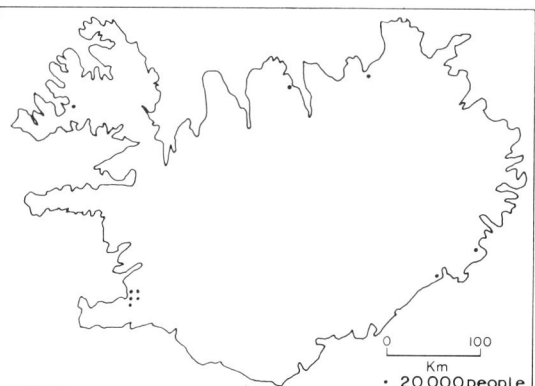

Fig. 7.3. Population density: Iceland.

employ millions of people. The rural population is generally of fairly high density too, for these lands are productive of cattle, fruit, vegetables, grains, tobacco and other crops. Only in the western coastlands of the United States are population densities similar to those of the east to be found, particularly in California, where lie the great urban sprawl of Los Angeles and the agriculturally productive Central Valley (page 194).

All the peoples of the world require water, food, shelter, clothing and fuel. Only very primitive people who live very simple lives are self-sufficient, that is, able to provide for themselves everything they need. Even some of these nowadays barter or trade goods with others who can provide items they require. Most people rely upon money with which to buy their necessities. To obtain money they work. The International Labour Office divides the work man does into various groups:

Agriculture, forestry, hunting and fishing
Extractive industries (e.g. mining, quarrying)
Manufacturing industries
Building
Electricity, gas, water and sanitary services
Commerce, banking, insurance and estate management
Transport, warehousing and communications
Services (e.g. health, education, entertainment, restaurants, hotels)

If you study this list you will see that you and your family rely, either directly or indirectly, on the workers in all these occupations. Your house or flat was constructed by builders. You have lighting, heating and sanitation in your home. Your furniture was made from timber cut by forestry workers and shaped by workers in a manufacturing industry before being transported to your house. It will be worthwhile for you to make a list showing, in as much detail as you can, how your life is made possible by the work of other people, using the table to help you. You can see that our lives are bound up with the lives of other people all over the world. Man is not self-sufficient. In working to earn money to provide his own necessities and luxuries, he helps to provide those of others. Yet despite the enormous variety of possible occupations, there are many parts of the world where men cannot find work. Unemployment is one of man's major problems.

There are other great problems of world population. There are those which arise in the provision of food, from people being crowded too closely together, from people speaking different languages and being of different race. Some of these we consider next.

Work to do

1. *Copy the table opposite. Then put each of the following occupations at the side of the heading to which it belongs: gold miner, cotton-mill worker, plumber, insurance agent, farmer, postman, bank manager, steel-maker, carpenter, estate agent, trawler captain.*

2. i) *Name three hot deserts of the world.*
   ii) *Explain what is meant by an oasis.*
   iii) *Explain why oases can be areas of dense population.*
   iv) *Name one oasis town.*

3. i) *Explain why mountains are areas of scanty population in temperate lands.*
   ii) *Explain why mountain areas in the tropics are more densely peopled.*

4. *Select any one densely peopled country, and say why so many people live there.*

5. i) *What is the chief cereal of South-East Asia?*
   ii) *Give three reasons why this crop is so important.*
   iii) *What cereal is mainly eaten by Europeans?*
   iv) *Do you think Europeans would starve if their grain crops failed?*

2. Overpopulation and underdevelopment

World Population
(in millions)

	1800	1850	1900	1950
Africa	90	95	120	198
Americas	25	59	144	321
Asia	602	749	937	1295
Europe	187	266	401	559
Oceania	2	2	6	13
(Australia and New Zealand)				

As you can see from the table, the population of the world is now increasing at a very rapid rate. It has been worked out that *every day* there are an additional 130 000 babies born. In earlier years many babies died at birth; with the advance of medicine not only do relatively few babies die but more and more people live to old age. Growing sufficient food for these vast populations is the greatest problem of the world today.

Of the total world land surface of some 146 million square km, approximately one-fifth is too mountainous or at high elevations, one-fifth is too cold, one-fifth is too dry and one-tenth has inadequate soils for cultivation. A little arithmetic shows this to be 70 per cent of the land surface. This means that approximately 30 per cent only has relief, soils and temperature suitable for cultivation. North America and Europe, including the U.S.S.R., have one-quarter of the world's population and three-quarters of the world's food. The other continents have three-quarters of the population and one-quarter of the food. It is with these continents that we are particularly concerned.

The populations which show the greatest rate of increase today

Standard of living: Chuck wagon: Arizona.
These people also move about with herds (page 198). List the more elaborate equipment and foodstuffs they possess.
(*U.S. Information Service*)

are those where the standard of living is low, as it is in South-East Asia. The standard of living might be described as the style in which people live. A high standard of living, familiar in Europe, North America and Australia, means that most people have a nutritious and varied diet, houses with running water and sanitation, and luxuries like radios, television sets, washing machines and cars. A low standard of living means that food is less plentiful and often short of protein or vitamins, houses generally lack running water and proper sanitation. Feeding and clothing a family may be difficult; money is very limited and there are no luxuries.

A low standard of living may occur among primitive peoples who rely on hunting and collecting for their food. For example, South American Indians like those of the Amazon Basin generally live far away from towns. They have no luxuries as we know them because their lives are simple. The bushmen of the Kalahari Desert have a low standard of living, because even surviving is difficult in desert conditions; they are forced to be nomads searching for food and water. The low standard of living of most people of South-East Asia, however, occurs because the land is over-populated. This means, put simply, that there are more people than farming and industry can support.

Study the table below. It gives the area and population of the countries of South-East Asia:

Country	Area 1000 square kilometres	Population in millions
Bangladesh	143	69
Burma	681	28
India	3268	550
Japan	372	105
Khmer Republic	182	6
Laos	237	3
Pakistan	804	59
Sri Lanka	65	13
Thailand	515	35
Vietnam	164	40

India, Pakistan, Japan and Sri Lanka are particularly densely peopled, and their populations are increasing rapidly each year. We will consider Japan later. Let us consider the other three countries. Most of the people are farmers. Their farms are very small, often of only one or two acres in scattered plots. In a good year, when rainfall has been plentiful, they may produce enough food for their usually large family with a small surplus to sell in town and village markets. In areas of unreliable rainfall, if the rains are less than is hoped harvests are very poor. The farmer finds it difficult or impossible to feed his own family, and has no surplus for sale. He is not only short of food but of money. The villagers, like the shoemaker, the potter, the weaver and the blacksmith, find no food in the markets; they are short of money too, for few can afford to buy their goods. The townspeople, who keep shops or work in factories, transport or offices, also find no food for sale. If a country cannot produce sufficient food it is necessary to buy food from elsewhere. Countries like our own pay for this food by making goods to sell abroad, and even in our industrial country we know that factory workers are always being urged to produce more for export. In India and Pakistan there are relatively few industrial workers compared with the total number of

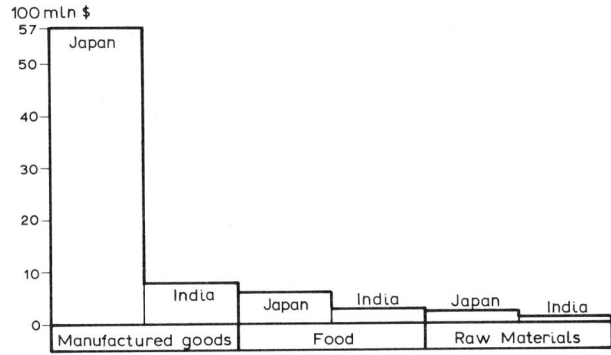

Fig. 7.4. Exports of India and Japan.

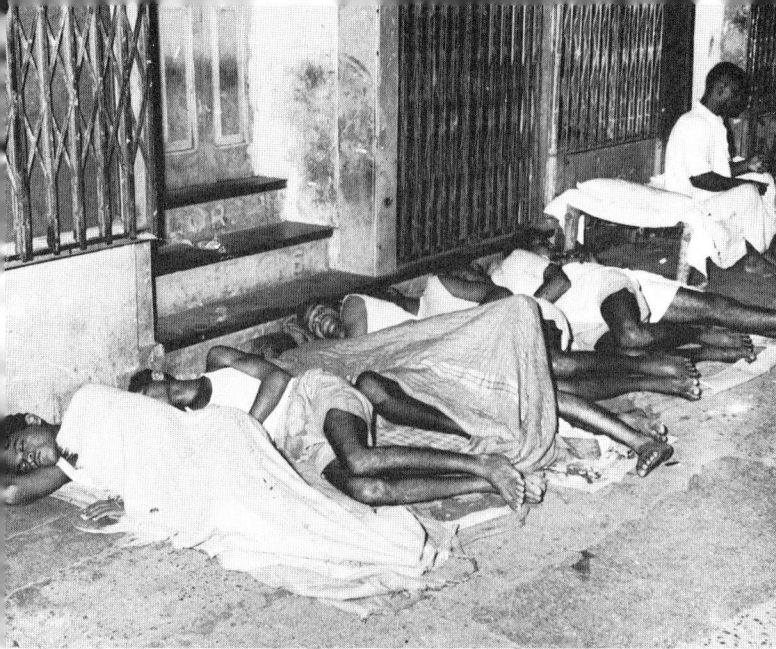

Poverty in India.
In this crowded country some people are homeless.
(*Camera Press*)

people, so that products for export are very limited (fig. 7.4). This means that imported food is scarce, since it cannot be paid for. Thus if home crops fail, famine results. There is insufficient food for millions of people. They may have but one meal a day, or even less. International organisations of all kinds send rice, wheat, milk powder and other foods in massive quantities free for distribution to the hungry people. Despite this, thousands die of starvation and from diseases resulting from malnutrition or insufficient nourishment.

There are many ways in which the governments of these countries are trying to remedy the situation. They are advising people to have smaller families, so that population increases less rapidly. They are trying to educate the farmers to be more efficient, so that by using better seed, more fertiliser and crop rotation heavier harvests will result. They have begun redistributing the scattered plots of land, so that each farmer's fields lie side by side and machinery may be used in them. They are extending irrigated areas by building dams and canals wherever there is adequate water available, and sinking more wells.

The governments are also trying to increase industrial productivity. This would not only give more for the home market and for export but is linked with improving agriculture. For example, the Colombo Plan of 1950 included a scheme for the construction of eight dams in the Damodar River valley, not only to provide water for irrigation but to provide hydro-electric power for factories producing fertiliser, farming machinery and other goods shown on the map (fig. 7.5). The development of such industries

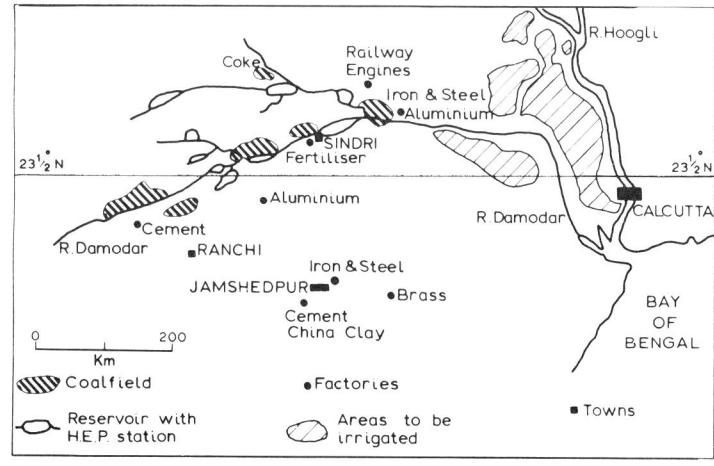

Fig. 7.5. The Damodar Valley scheme.

means employment for thousands of people and helps to raise the standard of living.

Countries where agriculture needs improving and there is need for industries are called 'underdeveloped'. Not all underdeveloped countries have vast populations; not all countries with vast populations are underdeveloped. Japan is an example of a small highly developed country with a large population whose standard of living is relatively high. This is due to great industries and intensive farming. Over 18 million people live in the seven towns shown on the map (fig. 7.6). They not only work in shipbuilding yards, iron and steel foundries, oil refineries, textile factories and food-processing plants, but in wholesale and retail trading, health services, public utilities and transport. Japanese factories are pro-

ducing more goods for export than formerly—cameras, radios, television sets, motor cycles and cars (fig. 7.4). Nevertheless, most people are farmers, and pressure on the land is great. The Government is reclaiming land from the sea, but this is possible only in limited areas. Food production is being increased by scientific methods. Only 5000 Japanese emigrate yearly, mainly to Brazil. They prefer their own country, and Japan still has the problem of overpopulation.

The Republic of China has an area of c. 10 million square km and an estimated population of 716 millions. Under her new government the people are learning to read and write, for they learn new ways of living more easily if educated. The small peasant holdings have been taken over and formed into large co-operative farms, where teams of farmers vie to produce the best crops, using mechanisation. Great rivers like the Hwang-Ho and the Yangtse Kiang (page 138) are being controlled so that they do not flood. New industries are springing up in Peking, Shanghai and other towns. Her coal, iron-ore, tin and copper mines are being developed. New railways and roads are extending transport. In these ways China, too, is trying to solve her problems of underdevelopment and overpopulation.

Africa and South America are also underdeveloped over large areas, but the pressure of population is at present less great than in monsoon lands. We shall see how man is trying to improve the savanna lands (page 179). We have already mentioned how shanty towns (page 113) sometimes grow up when people leave the land and go to cities. World organisations are sending technical advisers, machinery, doctors, medicine and teachers to these continents in the hope that the people will learn to use to the full the vast natural resources of their countries.

We have said that underdeveloped countries are those in which agriculture and industry need developing. This is a short description, for simplicity. By definition, according to U.N.O., underdeveloped countries are those which are in receipt of aid from the United States, Europe and the U.S.S.R. These countries are

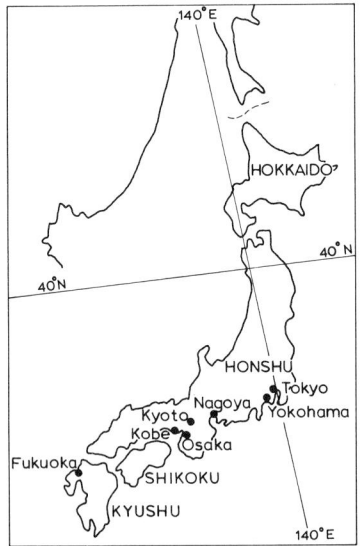

Fig. 7.6. Main industrial towns of Japan.

Algeria, Argentina, Bolivia, Brazil, Chile, Colombia, Zaire, Formosa, India, Indonesia, Iran, Israel, Kenya, Liberia, Libya, Mexico, Morocco, Pakistan, South Korea, Tanzania, Thailand, Tunisia, Turkey, United Arab Republic (Egypt), Venezuela, Yugoslavia and French oversea departments, together with the French franc area south of the Sahara.

Many countries in the Commonwealth receive the bulk of their aid direct from Great Britain through the Commonwealth Development and Welfare Fund, now over £200 000 000 per year.

Work to do

1. *Draw a pie (circle) diagram to show the information given in paragraph 2 (page 157). Scale: radius of circle = 5 cm (representing total land surface of the world). From the centre of the circle measure for mountainous land 72°, land too cold 72°, land too dry 72°, inadequate soils 36°, suitable for cultivation (the rest) 108°. Colour each section, and give your diagram a key and title.*

2. i) *Name three countries in monsoon Asia.*
   ii) *What is the chief food grown in this area?*
   iii) *Why is meat seldom eaten in this area?*
   iv) *Suggest how the farmers can produce more food on their farms.*
   v) *What is meant by 'insufficient nourishment'?*

3. i) *Why do harvests sometimes fail in South-East Asia?*
   ii) *When food is imported to overcome famine what do you think are the difficulties found in distributing it?*

4. *On a large outline map of the world, locate and name as many underdeveloped countries which are in receipt of aid from the United States, Europe and the U.S.S.R. as you can.*

5. *Write a sentence each to explain what is meant by:*
   i) *Overpopulation.*
   ii) *Underdevelopment.*
   iii) *High standard of living.*
   iv) *Low standard of living.*

3. Migration

Migration means movement from one place to another. Movement of peoples from their home country for the purpose of settling in another land is called emigration. In early days if people emigrated, their journeys overland or by sailing vessel across the seas were often long and dangerous. By the 1870s steamships were in use, and these not only shortened sea voyages, so making them cheaper, but made them safer. Early emigrants seldom returned to their native land; nowadays in general some three-quarters of the world's emigrants settle permanently in the country of their choice. The rest, with cheap and speedy transport now available, return home when they have saved enough money to do so.

Man has always attempted to leave lands in which life is difficult for those where he thinks conditions will be better. Migrations of people have been known throughout historic time, and they occur for various reasons. One reason has been the desire to escape persecution at home. You are all familiar with the sailing of the *Mayflower* in 1620 when the unpopular Puritans left England to colonise the land in America which they named New England. In 1684 persecution of the French Huguenots or Protestants caused thousands of them to flee across the Channel to England. Much more recently the Jewish people fled from Nazi Germany to find freedom in England, the United States and other countries. Many are brilliant scientists, doctors, writers and musicians whose talents are of great value to the countries which received them. After the great displacement of Jewish peoples during the Second World War the newly created state of Israel gave some of them a land in which they could settle.

Another reason for emigration is unemployment in the home country. Unemployment has been the chief cause of emigration from Europe. Many of those who left home were land workers from rural areas. At the beginning of the nineteenth century there was much emigration from the Highlands of Scotland, and even today the farmers or crofters of these areas often leave their homeland for Canada and Australia (fig. 7.7). Similar emigration

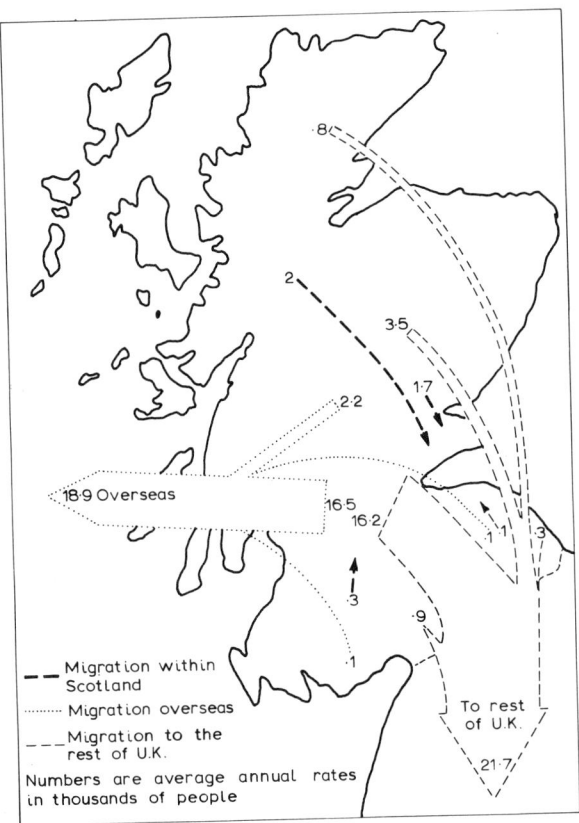

Fig. 7.7. Scottish migration 1961–67.

occurred from Germany and southern Italy, where large estates provided only seasonal work for the peasants. In 1820, after the Napoleonic wars, the price of corn fell so low in England that many farmers were ruined, and left to start new farms in North America (fig. 7.8). Another wave of emigration followed in 1870, when wheat was first imported from new wheatlands overseas, so that home production fell. Between 1850 and 1900 unemployment and famine caused more than 4 million Irish to leave their island, mainly for New York and Chicago, though many came to England, as they still do. Spanish, Portuguese and Italian farmworkers also emigrated in large numbers at this time, mainly to Argentina and Brazil, where the climates are warm like their own. This movement of rural people from lands which do not provide them with work continues today.

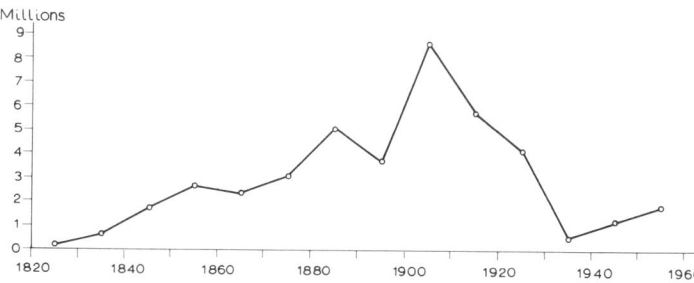

Fig. 7.8. Immigration into the United States.

Lack of employment also causes industrial workers to emigrate. During the Industrial Revolution the hand work of craftsmen was gradually replaced by the products of machines. These craftsmen or artisans were not necessarily poor, for many had worked hard and saved money, but their skills were no longer necessary, for machines could produce more goods in less time with fewer men. Thousands of British, Swiss, German and French artisans emigrated to the still largely unpopulated lands of North America. There they could either set up their own crafts again or work in factories as these were established. This migration of artisans continues; it is particularly noticeable in the movement of Scottish engineers and other shipyard workers from the Clyde estuary in times of depression, when orders for ships are few.

The discovery of gold has always attracted people. In 1849, for example, deposits of gold found in California brought thousands of prospectors, many of whom, unsuccessful in their search for the precious metal, remained to settle there. In 1851 rich alluvial gold deposits were found on what was then the southern shore of New South Wales. These deposits proved so valuable that thousands of people poured into the land; the new state of Victoria was created, and its population increased from 80 000 to 450 000 in ten years. A similar gold discovery in Western Australia in 1892 caused a further influx of immigrants, that is people entering the country. The discovery of other metals has also attracted immigrants. For example, many Cornish miners, unemployed after the local tin mines were abandoned, found work in the iron mines of Michigan, U.S.A., which opened in 1844, and in the gold mines of South Africa. There is no doubt, however, that the prospect of possibly making a fortune in goldfields attracts the largest number of people.

Not all emigration has been voluntary. You will remember from history that kings used to banish their enemies from their native land. Perhaps you can think of some examples. In the past in the countries which had colonies, like Britain, France, Spain and Portugal, the punishment for various crimes was transportation. The crimes committed were not necessarily very serious, but they resulted in the removal of the convicted person or convict to another land oversea for the rest of his life. Before the United States gained her independence from Britain in 1783 convicts were sometimes sent to our colonies there, but the most famous penal settlements were in Australia. The great city of Sydney developed from the early settlement of transported prisoners in 1788. Tasmania, too, had penal settlements. In 1815 the British Government in India sent convicts to the island of Mauritius, where they built up the sugar-cane plantations. Transportation was abolished as a punishment in our country in 1853.

Pressure of population has resulted in much emigration from Asia. Indians emigrated to Africa when labour was needed to establish plantations and to construct railways. Many railway workers, recruited by the British for work in Natal, Kenya, Tanganyika and South Africa, settled there after the railways were finished, and there are now thousands of Indians in these lands. Indians also began emigrating to the West Indies in 1840; large numbers still live in Trinidad, Jamaica and Guyana. They arrived in Fiji in 1874; today they number nearly 200 000 over half the population. The Chinese, too, have emigrated in large numbers. They adapt themselves very quickly to new surroundings, and are willing to take all sorts of jobs, sometimes as unskilled labourers, but often as traders, when they may become very wealthy. 60 000 went to the United States, particularly to California, 12 000 to Canada, 35 000 to Australia, 45 000 to Peru and 25 000 to Hawaii. Indeed, they worked so hard and lived on so little that they came to be regarded as a threat to white workers in some countries. In 1907 the State of California made severe laws restricting their entry, to be followed by similar laws in all the United States, Canada, Australia and New Zealand. Partly because of these laws, but also because some of them hope to return home, the majority of Chinese emigration, as fig. 7.9 shows, is within South-East Asia. The Japanese, too, are not permitted entry to some countries, and apart from the movement of over 1 million emigrants to Formosa, Korea and Manchuria before the Second World War, there are only about 25 000 Japanese in other lands, 15 000 of these living in Brazil.

Fig. 7.9. Chinese emigration in South-East Asia.

There are many problems created by emigrants. Sometimes the country they leave can ill spare them. For example, so many highly skilled scientists have left our country recently that the Government is worried about this 'brain drain'. Arriving immigrants are often poor, with few possessions; they are forced to find cheap housing. They may be hardworking, but often find only unskilled work with low wages, so that they cannot afford to paint and repair their homes, which may become very shabby and eventually form slums. If they work hard, save money and perhaps buy a new house or a car they may be regarded with jealousy. The immigrants do not always mix easily with the inhabitants of their new country. They may preserve their own language and customs, and be regarded with suspicion as 'foreigners'. Sometimes so many immigrants arrive that they outnumber the original population, who then fear that they will have difficulty in remaining in control of affairs. This is happening in Fiji. The state of Israel was created from land formerly occupied by Arabs, who resent being displaced by these hard-working immigrant Jewish

Islam in London: mosque at Southfields.
In what countries are buildings such as this normally found?
(*L. J. Long*)

peoples, whose religion is so different from their own. It is in countries like the United States, whose whole population is made up of emigrants from different countries, all wishing to speak English and become American citizens, where the situation is happiest.

Although we have spoken of our countrymen emigrating from our islands, we should also remember that many people from other lands have landed here as immigrants. It has been estimated that there are now 2 million non-native people living in the United Kingdom. These consist of some 700 000 Commonwealth citizens (300 000 Australians, Canadians, New Zealanders and South Africans, 300 000 West Indians and 100 000 Asians); 700 000 Irish from Eire; and 400 000 foreigners, including 100 000 Poles, 60 000 Italians and 40 000 Germans. The freedom of this country has always attracted people of other lands, and the figures have been rising rapidly.

165

Work to do

1. *Make a list of reasons why people emigrate.*

2. *Study the map of Scotland* (fig. 7.7) *and answer the questions:*

    i) *How many people migrated within Scotland, that is from one part to another?*
    ii) *From which parts of the country did they come? Where did they go to?*
    iii) *How many Scots migrated to England? Suggest reasons for their migration.*
    iv) *How many Scots emigrated overseas? Name some countries to which they went.*

3. *The following names are from a class register in a school in Mauritius, an island in the Indian Ocean. Study them and suggest the countries from which the childrens' families originally came.*

Abdulla, Shahid Nawab	*Chotnick, Ropchand*
Ah, Sen Lee Wun	*Chundoo, Abdul Alim*
Albert, Joseph Sydney	*De Senneville, Jean Michel*
Appiah, Daneswar	*Dindayal, Duleep Kumer*
Babajee, Janmajaye	*Gangoo, Mohammed*
Chasteau de Balyon, Bertrand	*Hung, Fong Yu*
Chen, Kai On Nan	*Hurree, Chandraneth*

4. *Study the map showing Chinese emigration* (fig. 7.9), *and make a list of the countries to which the Chinese have emigrated. Add any other countries to which they have emigrated (from the text), and give a reason why these have not been included on the map.*

5. *'It was not easy to make a start. In Naples, he had been a fruit vendor with a small pushcart. He used to buy his produce from the farmers who drifted into the city from the outlying districts, and he used to shove his pushcart all over the city, sometimes not getting home until nine or ten at night, but nonetheless providing a living for himself and his family. The living was poor, even by Italian standards; in Naples Salvatore Palumbo and his wife had lived in a slum. In America he moved directly to another city and another slum. He did not like the slum. In Italian, he said to his wife "I did not come to America to live in yet another slum . . ." You don't have to stay in a slum. If you have the will, determination and the ambition of a man like Salvatore Palumbo you can own a little house, and you can have your own fruit and vegetable store.'*

    i) *What nationality was Salvatore Palumbo?*
    ii) *Why did he leave his own country?*
    iii) *Why do you think he came to America?*
    iv) *Name the three qualities of character which enabled Salvatore to succeed.*
    v) *What did he end up owning?*

4. Race and language

You have all seen in the newspapers mention of colour prejudice and racial problems. You might say that this is a political matter, nothing to do with geography. This is quite true. Geography is not concerned with political opinions or the morals of being friendly or otherwise to other people. But geography studies people, where they live and where they move to, and it is only by having a clear idea of some of the facts about people that you can start to think for yourself about the problems involved. You will find that behind the difficulties that you read of in the papers are the simple geographical facts that people nowadays do not always stay in one place, but move about the world. Owing to various happenings in the past, and in the present, some of them have got mixed up, so that there are different groups of people living in the same area. When these areas are crowded, so that there is hardly room for all, difficulties arise and disagreements grow greater.

You will notice we have said 'groups of people' rather than races. To begin with you should remember that there are plenty of disagreements between groups of people, without their being of a different race. We must be very careful what we mean by race, for this is a word which can be used very vaguely. The people who study races are called anthropologists, and they try to classify the peoples of the earth into different groups. The first and obvious grouping is by colour of skin—white, yellow and brown (fig. 7.10). They try, too, to measure other features, such as shape of face, size and straightness of nose, height of cheekbones, and the colour and straightness of hair. This is all they attempt to do, but the work soon becomes very complicated, particularly where there is inter-marriage among fairly similar peoples. There is also a steady and slow change going on over much of the world, and peoples, through the years, migrate slowly from their original homeland to other places. One source of error in thinking about races arises when the word race is applied to a particular country.

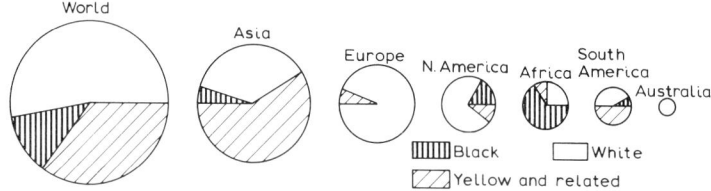

Fig. 7.10. Diagram of world races.

We can say, for example, that in general people of north-west Europe have white skins, fair hair and other similarities, but it is quite impossible to prove that there are clear distinguishing features of a Danish race or a Dutch race.

Let us see what we can safely say about racial groups. It is clear that most of the people of Europe are white-skinned, and the word Caucasian race is applied to these people, because they are believed to have come originally from the Caucasus Mountains. Many of the people of China and South-East Asia have yellow skins, and are called the Mongoloid race. Eskimos and the original Red Indians, both races which have almost disappeared, are believed to be branches of this. The people of Africa have brown skins and usually very curly dark hair, and are called the Negroid race. Once we go beyond this simple classification, complications arise. There are plenty of dark-skinned Africans in East Africa and the Sudan with straight noses. The brown-skinned people of India have straight hair and facial features very similar to those of Europeans. There are many people in southern Europe with swarthy skins and curly hair who look like rather pale Negroid peoples. In fact, you must be very careful when you use the word 'race' or speak of 'racial problems'. There are many difficulties in the present-day world between groups of peoples who live in the same country and do not get on with each other; some of these groups can be identified by the colour of their skins. It is partly because this particular difference is so readily visible that people speak of the colour problem or of racial problems.

We shall gain much more enlightenment—and be much nearer the study of geography—if we think of the problems caused by the mixing of peoples. Some of them have different coloured skins or shape of face. Much more important differences are in their ways of life, and quite a big part of this is their language. In many cases, though not all, the difficulties of living together and understanding each other are caused by speaking different languages.

As we live on an island, with clear-cut boundaries, we tend to think of our land frontier as matching that of our language. English is an official language, spoken by all. If we go to France we expect the people to speak French. But this is not entirely so. That part of France known as Alsace has belonged at intervals through history to Germanic peoples, and German speaking is common in Alsace. The map (fig. 7.11) shows you that Switzerland is a country which has four official languages—Romansch, German, French and Italian. Only some 50 000 people, roughly 1 per cent of the Swiss population, speak Romansch, a language which derives partly from Latin, a relict of the early Roman occupation of Switzerland. The later invasion of Switzerland by Germanic peoples brought the German language,

now spoken by about 70 per cent of the population. In the cantons (districts) of Valais, Neuchatel and Geneva, which became part of Switzerland in 1815, French is spoken. This is the language of some 20 per cent of the population. Italian is the language of the inhabitants of the canton of Ticino in the south. Many Swiss people speak two languages, with English as a common third language. Although there is no Swiss language, and official notices have to be printed in all four languages, the Swiss people form one united nation to which they are all proud to belong. In Belgium, established as a kingdom in 1839, there are two official languages (fig. 7.12). The peoples of north Belgium

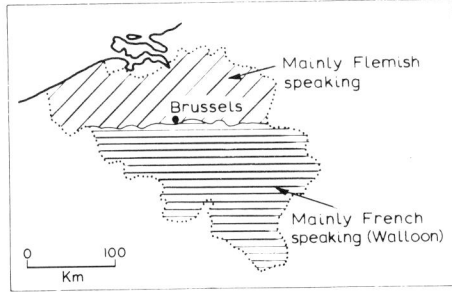

Fig. 7.12. The language division of Belgium.

mostly speak Flemish, a language which is a variant of Dutch. In the south French is generally spoken. These differences of language are strengthened by other differences; the Flemings are traditionally Catholic, those who speak French are generally Protestant. The two groups in general support different political parties. Thus there is a lack of the feeling of unity which binds the Swiss together as one nation; the Belgians are divided and the two language groups do not co-operate very easily.

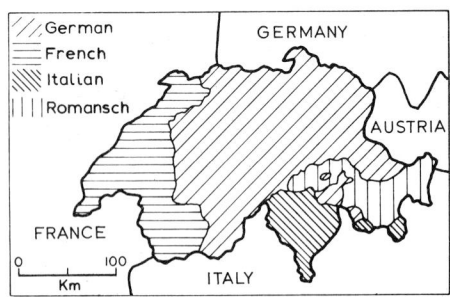

Fig. 7.11. The languages of Switzerland.

The Indian sub-continent well illustrates the difficulties arising from differences in language. Under British rule the official language was English, and this gave the possibility of a common tongue for millions of India's inhabitants, whose native languages are very numerous indeed. Soon after the independence of India in 1947 the Indian Government decided to abolish English and substitute Hindi. This decision is still resisted by the millions of Indians who speak other languages such as Gujarati, Marathi and Tamil. By contrast, in the United States, as the first settlers were English speaking, English has always been the dominant language, and the many waves of other people who entered the country were forced, of necessity, to adopt it.

There are many parts of the world where more than one group of people is living in the same country. Malaysia is a good example. The original inhabitants were Malays, a people who lived in most of the islands and peninsulas of South-East Asia. They are brown-skinned, though with some slight mongoloid facial features. They are generally regarded as good-natured, easy-going people, whose main occupation is farming and fishing. With the growth of rubber plantations and tin mining in the country, additional labour was needed, but the Malays were not attracted to rubber estates and tin workings. Large numbers of southern Tamil-speaking Indians came to live in the country, and they still do. Some keep shops, but most work for wages, producing rubber and tin. But Malaysia is also near the crowded country of China, and for many years some Chinese have been moving there. They are craftsmen and traders, and live mostly in towns as shopkeepers. Now only about one-third of the people of Malaysia are the original Malays. These three groups, who are only now beginning to live together as one country, could be said to be of different races, but their difficulties much more arise from their different languages and ways of life, as peasant farmers and fishermen, as rubber-tappers and as traders.

Perhaps the biggest problem of all today was originally caused by the mass movement of the coloured peoples of Africa during

Chinese shops in Singapore.
This scene with its open-fronted shops is typical of Singapore outside the central business district. What languages are shown?
(*Camera Press*)

the slave trade. For several centuries, until the abolition of the slave trade in the British Empire in 1833, millions of Africans, mostly from West Africa but also from the east, were forcibly transhipped to the West Indies and the southern part of the United States. They were used to provide cheap labour on the sugar and cotton plantations, and there are now millions of coloured people of African descent in the United States. What has happened since then in America is a matter for the historians and politicians to study. So far as geography is concerned, you should remember that until some thirty years ago most were still employed in the south, growing cotton and sometimes running small farms. In recent decades, however, cotton has become far less important, and there has been enormous expansion of the northern industrial cities of America. More and more coloured Americans—for they can no longer be called Africans—have moved to industrial towns in the north and on the Pacific coast. They are often unskilled workers, with low wages, who must therefore live in the poorest parts of the towns, and it is in these areas that disturbances have been occurring.

The situation in the West Indies was perhaps not quite so difficult. Besides the cotton and sugar of the early days, many other crops were grown, and the banana industry became particularly prosperous. Many other people settled in the West Indies, including British, French and again some Indians, who came to work on plantations. The islands are small, however, and the population grew rapidly. Since 1945 increasing numbers of West Indians have been emigrating to Britain, and as you probably know, they have come in such large numbers that the British Government has decided to restrict their entry. They have moved —often quite suddenly by air—from warm tropical islands, where most people live on farms or at least in country districts, to our rather cold and damp island, where jobs are to be found only in the towns. Naturally they have grouped together in particular parts of the big towns, such as London, Birmingham and Leeds. As they had not much money, and were not often skilled workers,

they had to live in the cheapest houses. Can you wonder in such circumstances that they find life strange and difficult sometimes?

We have mentioned only a few parts of the world where there are differences between people. There are many other parts where problems exist. It is important that you remember that problems arise not because people have different coloured skins but because their ways of life differ. Geography helps us to understand why people live in different ways, and to understand their problems. Basically all mankind has the same problem, that of procuring food, shelter, clothing and work.

Work to do

1. *Describe carefully the facial features of:* (a) *Caucasian;* (b) *Mongoloid;* (c) *Negroid peoples. Why is this not easy to do?*

2. *Make a list of any occupations you know to be followed by immigrants in our country.*

3. *Imagine yourself an immigrant in a strange country such as Africa or South America. Write an essay to describe the difficulties you would have to face.*

4.   i) *From which countries did the forefathers of the present coloured Americans come?*
    ii) *Why have many coloured Americans left the southern United States?*
   iii) *Where have they gone to?*
   iv) *What problems does their presence create?*

5. *Why is emigration much easier today than it used to be? Why is it also in some ways more difficult?*

Game on the African savanna. ▶
(*J. Allan Cash*)

Chapter 8

PROBLEMS OF REGIONS

1. Equatorial forests

If you study the picture of the equatorial forest you will see in the foreground dense undergrowth of bushes and fern. This undergrowth grows because, as you can see, sunlight penetrates between the trees. In the background of the picture the trees are all so closely packed that it is difficult to pick out any individual tree, and the darkness shows that no sunlight can reach the ground, so there is no undergrowth. The big trees in the picture, of which part of the trunk only is visible, are very tall, probably about 45 m high. The base of the trunk is wider at ground level to form a buttress to help support its great height. There are over 2000 kinds of trees in the equatorial forests, but you need only remember a few names, such as mahogany, silk-cotton, rubber and balsa. The tall trees form a canopy over smaller trees, among which may be found the banana and the cacao tree. Growing among all the trees are great creepers or lianes. Brilliantly coloured flowers, such as orchids, grow in leaves which have lodged in the cracks of branches and form a rich soil.

If you visited the forest the first thing you would notice would be the smell of wet and rotting vegetation. The trees do not lose their leaves all in one season, as ours generally do, but shed them gradually, so that there are always new leaves growing and old leaves rotting on the wet ground. The second thing you would notice would be the noise. The forests abound with chattering birds, such as parrots, with monkeys and other animals, with frogs and hundreds of different insects. You would also soon be aware of the humidity of the forest, that is its damp heat, which would make you perspire, so that your clothes would stick to your body and you would long for a cold shower.

Equatorial Forest: Brazil. ▶
(*Camera Press*)

172

In the equatorial forest the daily temperature, throughout the year, averages 27°C. This high temperature is considered heat-wave temperature in Britain. If you think that such heat might be a pleasant change, remember that, whereas London has an annual rainfall of 610 mm, these equatorial areas are also some of the wettest of the world, and receive 2000–2500 mm of rain a year. High temperatures mean rapid evaporation, so that the air is always humid, charged with moisture. The constant heat causes this hot damp air to rise so that massive thunderclouds form, and rain falls in torrents daily. It is, in fact, the perpetual heat and wet which cause the great forests to grow.

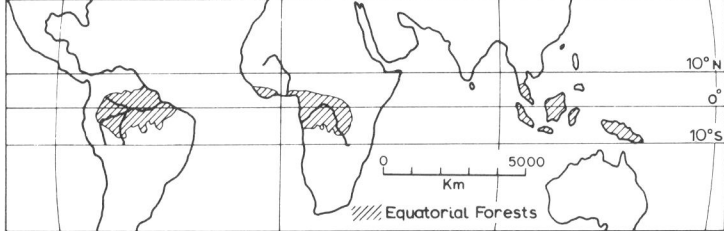

Fig. 8.1. Equatorial forest lands.

The map shows you the location of the equatorial forests in the world (fig. 8.1). The least developed of them is that of the Amazon Basin in South America. Despite its vast size, relatively few people live here. The native tribes of Amazonian Indians live by fishing in the River Amazon and its many tributaries, by hunting animals such as the monkey, and by collecting roots and fruits like bananas. Some of them live in huts built on stilts on the river banks. Some of them burn clearings in the forest and grow cassava. This tropical plant has a large tuber or root which has to be dug up, shredded, washed and pressed to remove the prussic acid it contains. After preparation it forms a flour for cooking. You may have eaten cassava in the form of tapioca pudding.

Felling mahogany: Nigeria.
(*United Africa Company*)

The obvious occupation in this area would at first sight appear to be the felling of the trees to sell as timber. There are, however, many difficulties. Manual labour is not easy in such a hot, humid climate. Trees of the same kind grow far apart from each other.

Indeed, the men who cut mahogany are known as mahogany hunters, for they have to search for trees to cut. The dense undergrowth makes it difficult to hack a way to the tree. The trees give very hard wood, and are difficult to fell. Before felling it is necessary to axe all the creepers away and clear a space so that the tree can fall to the ground. There are few roads or railways, so that the rivers are the only means of transport. This means that, in general, only areas near to the rivers can be cut for timber. Even when the huge trunks have been hauled to the river's edge, most of them are too heavy to float. They have to be placed on rafts of light wood, such as balsa, to be floated downstream to river ports such as Manáus or Belem for final export.

Besides offering opportunities for cutting timber, the forest provides other materials which can be collected or gathered. Latex, the sap of the rubber tree (page 73), is collected, so that some 40 000 tons of rubber are yearly available for use in Brazil. Chicle sap is collected in the same way to be made into chewing gum in the United States. Brazil nuts are gathered. Kapok, the fine cotton-wool type of fibre which surrounds the seeds of the silk-cotton tree, is also gathered. It is used all over the world to stuff cushions. The nuts of the tagua tree are collected to be made into buttons such as you may be wearing. In the west, around Iquitos, bark is stripped from cinchona trees to produce quinine, a drug useful against malaria, while coca leaves give cocaine, a pain-killing drug which your dentist may use.

There are many countries interested in developing the Amazon Basin. Since the greatest part of this enormous area lies within Brazil, her government has the greatest interest, but the Guianas, Venezuela, Colombia, Peru and Bolivia all have some part of the forest within their boundaries. Henry Ford, the car manufacturer, tried to establish rubber plantations in eastern Brazil, for the wild rubber trees, like all others, lie far apart, so that latex collecting is difficult and slow. The plantations were not successful, for the skilled labour necessary to look after the trees was not available. There are attempts to produce crops of cocoa, sugar, cotton and maize near Manáus and Belem. The Peruvian Government is encouraging some of its highland tribes to settle in the lower lands of the Amazon Basin, where it is hoped that they will find farming easier.

The equatorial forests of Africa lie within the Congo Basin and along the coast of West Africa, where they are fringed by mangrove swamps. These forests are more open than those of South America, and cultivation is somewhat easier. There are still a few tribes who live by hunting and collecting, but most Africans practise cultivation (page 70) near the rivers, growing yams, sweet potatoes and sometimes rice. There are more small settlements whose inhabitants grow fruits such as bananas, guavas and mangoes, together with other crops such as sugar-cane, beans, yams and even tobacco. The most important tree in the forest is the oil palm, which not only grows wild but is now cultivated in West Africa, particularly on the east of the Niger delta. The tree grows to a height of some 10 m, with a cluster of plum-coloured fruits called a fruiting head at the top. Fig. 8.2 shows you the

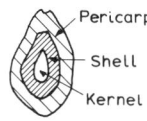

Fig. 8.2. Section of oil-palm fruit.

fruit or oil-palm nut. The outer covering (pericarp) is pulped locally to give oil for cooking. The kernels are usually exported to both North America and Europe, where the oil is extracted for use in making margarine, soap and paint. West Africa supplies 60 per cent of the palm-oil exports of the world, much of which comes from Nigeria. The other important tree is the cacao tree, the pod of which contains seeds or beans which are ground to give cocoa.

Palm oil fruit: Nigeria.
In the foreground is the fruiting head. Note the plantation of oil palms
in the background.
(*United Africa Company*)

The beans are dried in the sun before export to Britain, the rest of Europe and North America. The most important cocoa area lies in Ghana.

The equatorial lands of Malaysia and Indonesia contrast greatly with the sparsely peopled forest lands of the Amazon Basin. They experience high temperatures and heavy rainfall; they are also open to the influence of monsoon winds. You will see from the map that access is easy by sea, and that penetration inland involves only short distances. As a result of these and other factors, these areas are in parts very densely peopled. Java, for example, is about the same size as England, but has 15 million more people. Its high population density is partly the result of fertile volcanic soils which allow high yields, and the influence of the Dutch who originally took possession of the island. They introduced intensive garden methods of farming, grew sugar, spices, rubber, tea, coffee and cinchona in plantations, and developed roads and railways. Java is a beautiful island, with many villages of small homesteads, each with its plots of land surrounded by fruit trees. Sumatra is much larger than Java, but has more mountains and less fertile soil, so that the Dutch development of tea and rubber plantations on the hill slopes, and coconuts, cacao and oil palm on the low ground was more limited. We have already noted the importance of rubber in Malaysia (page 73). Rice is the chief grain crop of the whole area.

The problems of the area are not those of the other equatorial forest lands; they are the problems of overpopulation.

General lack of transport other than the great rivers, the climate and lack of labour all contribute to the under-development of the other equatorial forest areas, those of the Amazon and Congo Basins. The Amazonian Indians, and many Africans, suffer from diets lacking in protein and vitamins, so that they lack energy. The constant heat, constant humidity and constant rainfall are not attractive aspects of weather, and it is difficult to recruit labour from elsewhere. The forests are unhealthy; malaria and other fevers are common. There are few minerals to encourage settlement; there has been little oil discovered, though it has been sought. It does not seem likely that the Amazon Basin can yet support a large population. The Congo Basin, which has been open to European influence for many years, appears to have a more prosperous future.

Work to do

1. *Draw a sketch copied from the photograph on page 172. Label in the correct places: (a) tall tree; (b) smaller trees; (c) dense forest; (d) sunlight; (e) undergrowth; (f) liane or creeper. Give your sketch a title.*

2. *Draw two columns, heading the first Products and the second Uses. In the first column list all the products of the Amazon Basin, and write down the use made of them in the second.*

3. *Write a paragraph on why lumbering is difficult in equatorial forests.*

4. *The following are the temperature and rainfall figures for Georgetown (7° N., 58° W.).*

	Jan.	Feb.	March	Apr.	May	June
Temperature, °C	26	26	26	27	27	26
Rainfall, mm	200	117	183	152	282	297

	July	Aug.	Sept.	Oct.	Nov.	Dec.
Temperature, °C	26	27	28	28	27	27
Rainfall, mm	251	165	79	74	170	282

   i) *Draw graphs to represent these figures. Put a title, stating where Georgetown is.*
   ii) *What is the difference between the highest and lowest temperature? This is the annual range of temperature. Is it high or low?*
   iii) *There are no seasons as we know them in the equatorial forests. Why not?*
   iv) *What is the year's total rainfall in millimetres? What is London's total annual rainfall?*

5. *Write a paragraph describing all the dangers of the equatorial forests.*

2. Savanna lands

'Come on safari to East Africa,' says an advertisement, 'see majestic lions and graceful giraffes, rhinoceros and gnus, elephants and eland.' A safari is a hunting trip. In earlier days hunters with spears or rifles killed thousands of the wild animals. Nowadays many game reserves have been established, and weapons are not permitted (fig. 8.3). Visitors take cameras instead; it requires as much skill and patience to take a close-up of a wild animal as to use it as a target for bullets.

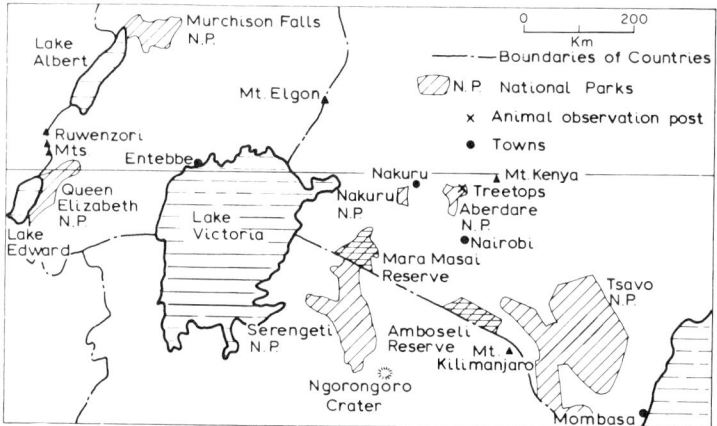

Fig. 8.3. Safari areas in East Africa.

The animals mentioned, and many others, live in the savannas of Africa. Lions, leopards, cheetahs and wild dogs are carnivores, that is they live on the flesh of other animals which they hunt and kill. Elephants, rhinoceros, giraffes, zebras, gnus, eland and other types of deer are herbivores, living on leaves, grass and other vegetation. Many of these animals are protectively stippled, spotted or striped so that they merge into the surrounding foliage, and rely on their speed to escape the hunting carnivores. Hyenas

and jackals are scavengers; if sufficiently hungry they may attack, but normally they eat what the carnivores leave of their kill. Another scavenger of the savannas is the vulture, a large, unattractive bird with a strong, hooked beak and big claws.

Safaris take place in the dry season of the savannas. In this rainless season the tall, coarse grasses dry up, their stalks becoming stiff and hard. The deciduous trees which grow among them are specially adapted to withstand heat and drought, for they have long roots or may store water in their trunks, as does the huge baobab tree, but in the dry season they shed their leaves. The herbivores then move to the river banks or lake shores, where there is not only water but tender young shoots and leaves are growing. They are followed by the carnivores and scavengers. The safari leaders know all the haunts of the animals, and visitors have the advantage of dry, sunny weather for their photography and travel.

Temperatures become highest just before the rains start. The torrential convectional rainstorms, often with violent thunder and lightning, cause a sudden, great change in the savanna landscape. The rivers swell and may flood, the unsurfaced roads and tracks become impassably muddy, the trees burst into leaf, the grass grows again rapidly. The wild animals, sure of plenty of water and food, can roam again more widely.

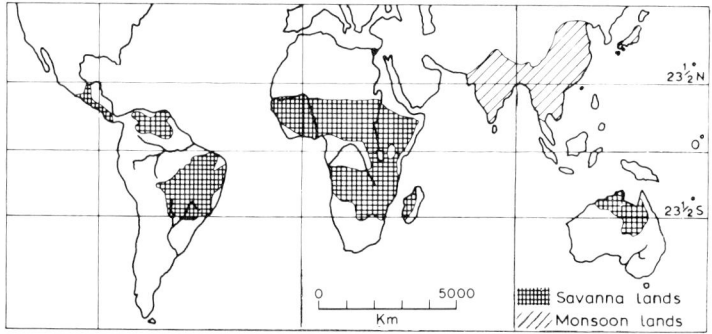

Fig. 8.4. Savanna and monsoon lands.

The savanna of Africa, as fig. 8.4 shows, covers a vast area. In the north near the Sahara Desert even the wet season rains are limited, so that the grass grows in tufts and there are many thorn bushes. As rainfall increases farther south the thorn bushes disappear, the grass grows taller and more luxuriant, and there are more trees. The poorer savannas of the north are called 'dry' savannas, to distinguish them from the rich parkland or 'wet' savannas of the equatorial plateaux.

Since the main vegetation of the savannas is grass, it is natural cattle country. African herdsmen have grazed large numbers of cattle over the areas for centuries. Cattle are regarded as the wealth of the tribe, and given as dowries in marriage. They are often unhealthy, and become very thin in the dry season when pasture is scarce. To provide grass for them at this time the herdsmen set fire to the dried, yellow grass stalks. These flare up and burn out rapidly, leaving any young grass shoots close to the ground exposed for the cattle. Unfortunately many trees are also destroyed in such fires, and do not grow again. Nowadays great efforts are being made to persuade the Africans to keep fewer animals in a more healthy condition, so that the land is not overgrazed. Most Africans now grow crops in the wet season which can be used to provide cattle food in the dry season. These crop lands are usually just patches of land near the village, and in the dry season the cattle wander over them freely. Every few years different patches are used, a system known as bush fallowing. Grass, weeds and shrubs grow up again, and the land recovers some, but not much, fertility (fig. 3.8).

The real problem arising from overgrazing the land and leaving it waste after bush fallowing is soil erosion. When the land is left bare of vegetation the heavy rainstorms of the wet season wash away the soil, especially if the surface of the land is sloping or hilly (page 9). Nothing will grow on the bare rock which is left. With the help of international organisations, present African governments are trying to stop soil erosion by encouraging modern methods of cattle farming and crop raising. In north

Savanna homestead: Tanzania.
(a) Study this picture and fig. 3.8, to write your own notes on a homestead in the African savanna. (b) Compare the vegetation in this picture with the savanna shown on page 180.
(*J. Allan Cash*)

Nigeria, for example, control of burning and grazing has produced improved grass cover, particularly where the land has been sown with special grass seed. Sections of 400 ha have been fenced and divided into five parts to be grazed in turn. Trees have been planted in river valleys; their roots check the rapid run-off of rain and hold the soil in place. Dams have been built near the sources with special grass seed. Sections of 400 ha have been fenced and irrigation during the dry season. Farmers have been given plots of land so that they can grow crops in rotation, using fertilisers. In northern Rhodesia, too, Africans have been given holdings up to

6 ha in size. The government clears 1·5 ha for each farmer, gives him ploughs and a harrow, four oxen and two cows, and lends him up to £150 in cash repayable within ten years. The farmer has to follow a special four-year rotation of crops, including beans, ground nuts (peanuts) and maize. Many such farmers are already producing maize, vegetables, milk and eggs for sale. In Tanzania, on the lower slopes of Mount Kilimanjaro, the average holding is 1·5 ha, which will grow about 200 banana plants and 600–800 coffee bushes. Bananas here supply food and beer and are fed to cattle. The coffee is sold. Yams, maize, onions, cabbages, sugar-cane and pineapples are also grown on these farms. Yet the savanna lands are so vast that it will be many years before the great problem of soil conservation can be solved.

Parts of the highland areas of the east and south of the African savannas are farmed by Europeans, many of whom have lived in Africa for several generations. Most of the savanna lies between the Tropics, but the height of much of the plateau land makes temperatures mild enough for permanent settlement by white people. In Kenya, for example, coffee and tea are grown for export, while many farmers keep fine, healthy sheep and cattle. In Rhodesia tobacco is an important export, while cane sugar and maize are also grown. Europeans are also training the Africans to use similar modern methods of farming. Thus it can be seen that the savanna lands can produce a variety of crops, while raising stock is still important over large areas.

Stock raising is particularly important in the savannas of South America and Australia. The savanna of South America lies in two areas, known as the campos of the Brazilian Highlands and the llanos of the Orinoco Basin. In the campos, cooler because it is a plateau land, there are some very large cattle estates supplying towns in the eastern coastal areas, such as Santos, São Paulo and Rio de Janeiro. In the llanos cattle raising is hampered by heat and disease, but ranchers are trying to improve their stock and are building dams to hold up the river waters for use in the dry season.

Cattle in the savanna: Queensland.
The tree in the foreground is a eucalyptus or 'gum' tree common over most of Australia.
(*Australian Information Bureau*)

In the Australian savanna are vast ranches called cattle stations varying from 2500 to 12500 square km in area. Here the lack of water in the dry season has been overcome by the boring of many artesian wells (page 38). To get the cattle to the coast for export they may be 'over-landed', that is driven on the hoof along tracks known as stock routes. Nowadays in some places cattle are slaughtered and the beef is flown to the ports. 'Air Beef' flying is very expensive, so in most areas 'road-trains' are used, each of which consists of two trailers pulled by a huge lorry, and carrying between fifty and eighty cattle. In the west these road-trains go to Derby, Broome or Meekatharra; from central cattle stations they

Road train: Northern Territory, Australia.
Compare this picture with the previous one, to remind yourself of two
types of savanna.
(*Australian Information Bureau*)

are driven to Mount Isa in Queensland for railway transport east, or go south by the one road, the Stuart Highway, which leads to Alice Springs, and thence by rail to Adelaide.

In Australia the savannas are less subject to soil erosion because they have not been overgrazed or improperly farmed.

There are few carnivores other than the dingo (wild dog), hyenas and jackals; the tsetse fly is not found (page 15). It is in Africa, where the savannas are most extensive, that the large-scale problems of conserving the soil, providing water in the dry season, farming scientifically and raising healthy stock must be solved.

Work to do

1. i) *What do you understand by savanna vegetation?*
   ii) *Describe the savanna:* (a) *in the wet season;* (b) *in the dry season.*
   iii) *Describe the life of a cattle herder in the West African savanna, explaining the usefulness of his cattle.*

2. i) *On an outline map of Australia, shade in lightly the areas of savanna as shown in fig. 8.4.*
   ii) *Mark in the towns named on page 179, using your atlas to help you.*
   iii) *Add the cattle routes mentioned in the text.*
   iv) *Give your map a title.*

3. *Imagine you are spending a day on safari. Write an essay describing all that you see and do, including information about the weather.*

4. *The following are the temperature and rainfall figures for Salisbury, Rhodesia:*

	Jan.	Feb.	March	Apr.	May	June
Temperature, °C	*20·6*	*20*	*19·4*	*18·3*	*15·6*	*13·9*
Rainfall, mm	*190*	*203*	*127*	*20*	*5*	*—*

	July	Aug.	Sept.	Oct.	Nov.	Dec.
Temperature, °C	*13·3*	*16·1*	*18·3*	*20·6*	*21·1*	*20·6*
Rainfall, mm	*—*	*—*	*13*	*30*	*99*	*152*

   i) *Draw graphs to represent these figures.*
   ii) *Is Salisbury north or south of the equator? How do you know?*
   iii) *Write down the months which form the dry season. What do you notice about temperatures during the dry season?*
   iv) *Which months would form the dry season on the opposite side of the equator?*

5. *Study the map* (fig. 8.3) *of safari areas in East Africa, then answer the questions:*

   i) *How many national parks are shown?*
   ii) *Name the countries shown on the map.*
   iii) *Name the sea in the south-east of the area shown.*
   iv) *Work out the area of Amboselli Reserve.*
   v) *What is meant by a reserve?*
   vi) *Find Mts. Kilimanjaro, Elgon and Kenya in your atlas. Write down their heights.*
   vii) *What do you think is meant by Ngorongoro Crater?*

181

3. Monsoon lands

A monsoon is a seasonal wind. A seasonal wind blows from the same direction for many weeks at a time. When the wind blows from the sea it brings with it clouds which produce rain. When the wind blows from the land it is a dry wind, so that few clouds form. In South-East Asia, from India to Japan, these seasonal winds are such a striking feature of the climate that we can conveniently consider these lands together, though there are considerable differences between them (fig. 8.4).

Large areas of these lands lie about the Tropic of Cancer, so you would expect temperatures to be high all the year round. Let us take India, Bangladesh and Pakistan, which form the Indian sub-continent. The lowlands are very hot, but the central plateau of India, the Deccan, has slightly lower temperatures, while in the mountains of the Himalayas, to the north of the Tropic of Cancer, the very many high peaks are always covered with snow and ice. Apart from the severe winters of these mountains, the Indian sub-continent does not experience winter; temperatures are at their lowest from mid-October to mid-March, and this is called the cool season. From mid-March to mid-June the weather grows hotter and hotter, the sun beats down, no clouds form in the sky, the plants shrivel and wither, many trees lose their leaves and streams dry up. This time is called the hot season. Towards the middle of June, when the heat is greatest, great banks of dark cloud appear, particularly along the west coast and over the Ganges delta area. Quite suddenly great storms break, and the rain deluges down. This is the rainy season. It lasts, with gradually decreasing rainfall, till mid-October. The clouds and torrential rainfall cool the air to some extent, but humidity is high, so that man finds this season particularly exhausting. As in the savannas, the whole landscape may change as vegetation, crops and rivers spring to life.

As the maps show (figs. 8.5 and 8.6), not all the Indian sub-

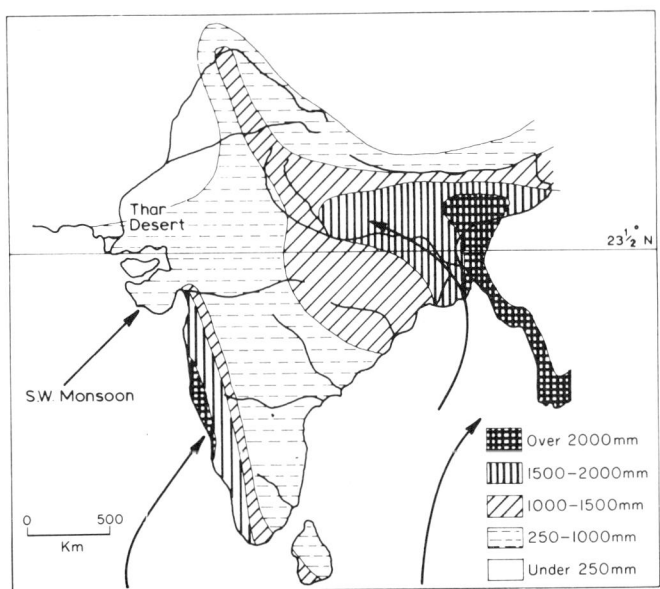

Fig. 8.5. Rainfall in India (May–October).

continent experiences heavy monsoon rainfall. The west coast of India, where the Western Ghats steeply edge the Deccan, receives over 2000 mm of rain; so does the area inland from the mouth of the Ganges, where the winds are forced to rise over the foothills of the Himalayas. In these foothills the rainfall may be over 10 000 mm, the highest in the world. Farther inland the rainfall is less, and since evaporation is high because temperatures are high, rainfall of less than 1000 mm may be scarcely adequate for farming.

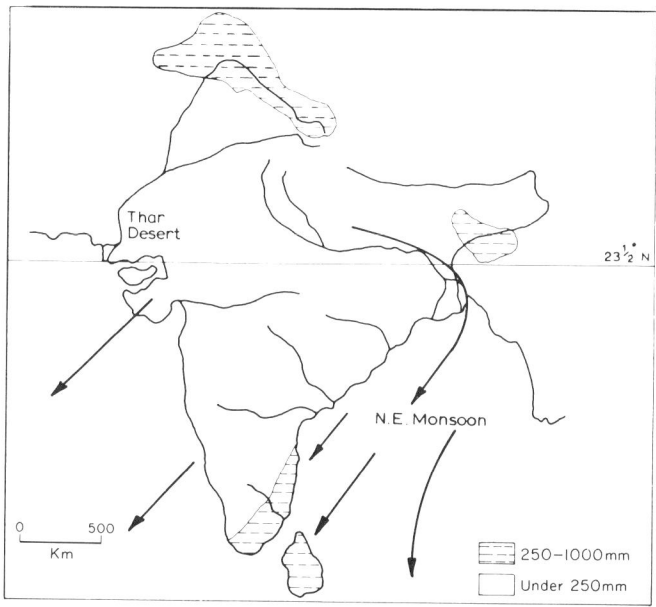

Fig. 8.6. Rainfall in India (November–April).

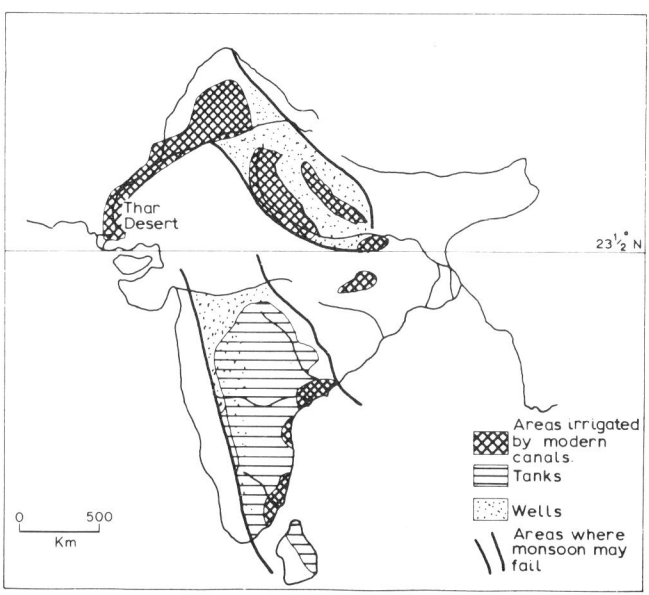

Fig. 8.7. Irrigation in the Indian sub-continent.

The lowlands of the Indus valley receive very little rainfall, and indeed without the snow-fed water of this great river and its tributaries for irrigation purposes the area would be desert, like the Thar region. In some years only limited storms of monsoon rain reach the central and eastern Deccan or the upper Ganges valley, and these are the areas of unreliable rainfall where crops may fail and famine may occur (page 159).

One of the problems of people living in these monsoon lands, therefore, is the unreliability of the rainfall. Most of them live by farming, and much of the farming relies on irrigation. The map (fig. 8.7) shows the main irrigated areas of India and Pakistan. In the Indus valley large steel and concrete dams hold back the river water in reservoirs from which water is fed to canals and so to the fields. In the Ganges valley wells are sunk into the alluvium, so that underground water is tapped. Over much of the Deccan, often where deep, narrow valleys occur, river and stream water is held back with small dams of earth and brushwood to form ponds called tanks (fig. 8.8). The danger of famine is greatest when the wells fail or the tanks dry up.

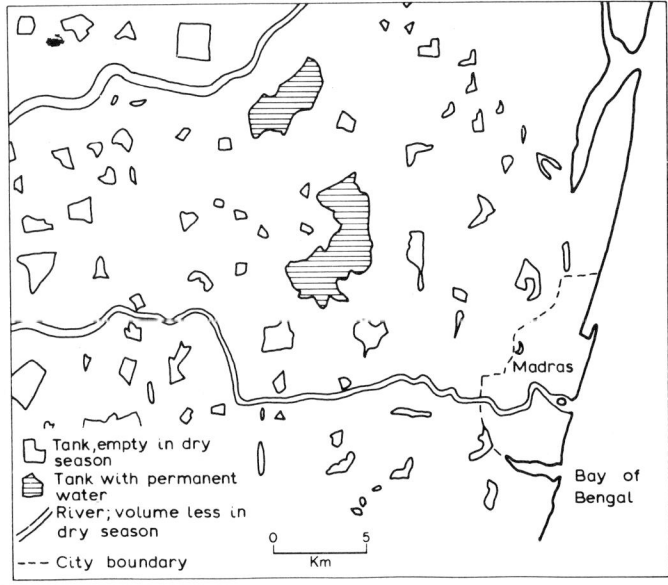

Fig. 8.8. Tanks on the Deccan.

A typical monsoon climate of the west coast of India can be summed up by the rainfall and temperature figures for Bombay:

Bombay	Jan.	Feb.	March	Apr.	May	June
Temperature, °C	24·4	24·4	26·7	28·3	30	28·9
Rainfall, mm	3	3	—	—	18	505

Bombay	July	Aug.	Sept.	Oct.	Nov.	Dec.
Temperature, °C	27·2	27·2	27·2	27·8	27·2	25
Rainfall, mm	610	368	269	48	10	—

Japan, too, has a monsoon climate, so let us look at the rainfall and temperature figures for Osaka, a port on Honshu, the main island of Japan:

Osaka	Jan.	Feb.	March	Apr.	May	June
Temperature, °C	3·9	3·9	7·2	13·3	17·8	22·2
Rainfall, mm	56	91	109	150	160	208

Osaka	July	Aug.	Sept.	Oct.	Nov.	Dec.
Temperature, °C	26·1	27·2	23·3	16·7	13·9	6·1
Rainfall, mm	180	190	183	109	109	58

Since Osaka lies at latitude 35° N., temperatures are different, and Japan has four seasons as we do, not three as in India. The distribution of rainfall is also very different, for the Japanese islands lie in a maritime situation between the Sea of Japan and the Pacific Ocean. The main rains of Osaka fall in the same months as those of the rainy season of Bombay, but cyclones which develop in winter pick up some moisture from the sea, so that Osaka has no months without rainfall. Japan is a mountainous country with many snow-fed rivers and streams, so that there is plentiful water for irrigation if it is needed.

There is an enormous variety of plants in the monsoon lands, so great that it is better not to speak of 'monsoon vegetation'. These lands spread from hot to cooler areas, include wet and dry climates, and many mountains. On the tropical lowlands, which have over 2000 mm of rain, the dense forests are little different from those of equatorial lands. The very word 'jungle' comes from India, meaning these tangled, impenetrable forests. In the drier parts, particularly of India, Burma and Thailand, the forest is more open, and the leaves fall in the dry season. There are many patches of bamboo, which grows quickly. Its strong light stem is very useful for making houses, furniture and other things. You may have heard of bamboo shoots as a Chinese dish. An important tree of the monsoon forests of Burma is teak. This is valuable because its wood is hard and resistant to waterlogging, so that it can be used for docks, ships' decks, boats and garden furniture. Another tree is the sal, or sandalwood, which has a pleasant scent and is used to line cupboards, or for trunks and caskets. Over much of China the climate is wet enough for evergreens to grow. The rhododendron and the orange both came originally from this country.

In Japan and other lands where the monsoon climate is generally cooler the forest is different. It is still mainly deciduous forest, but the trees lose their leaves in winter as ours do. Maple, birch, beech, poplar and oak are the most common trees. The mountains of Japan, like those of the Himalayas and parts of China, have deciduous trees on their lower slopes, but coniferous trees, such as cypress, cedar and pine, appear on the higher slopes where altitude makes temperatures colder.

It must be remembered, however, that over vast areas of monsoon Asia the natural vegetation has disappeared. It has been destroyed by man. Man has felled trees to provide timber for building and wood for fuel, to clear land for settlement and for farming. Although the more inaccessible mountain slopes are still heavily forested, the lower slopes are often cleared to make terraces for cultivation. Indeed, so much forest has been destroyed that lack of timber is a problem, for other forms of fuel are costly, and most people of South-East Asia cannot afford them. In tropical lands heating for houses is not necessary, but food must be cooked. Often animal dung is dried and used for fuel. Since this could be used to restore fertility to the soil, farming suffers, for artificial fertilisers to replace it also are costly. Bamboo often replaces wood as fencing. Man has also destroyed grassland areas by overgrazing and by growing crops, so that it is difficult to find food for animals, especially during the rainless months. Most farmers keep oxen or water-buffaloes to pull ploughs, tread grain for threshing and for transport. Such stock may be fed grain husks and stalks, crushed sugar cane, pea-pods, carrot tops and even leaves; the animals are often thin.

The monsoon lands are very densely populated (page 154);

most of the people are farmers. The chief crop we associate with farming in South-East Asia is rice, for more than one quarter of the cultivated land in India and south China, and more than half in Japan, consists of ricefields. Lowland rice is the main crop on coastal plains, flood plains, deltas and on terraces cut in the lower mountain slopes. Some varieties of rice grains will not start to grow in temperatures below 21 °C, and need higher temperatures during the four months' growing season. The picture shows the

Intensive cultivation: Japan.
Oiled paper is being used to cover these nursery beds of young rice to produce additional warmth in spring. Notice the irrigation channels in the foreground.
(*J. Allan Cash*)

Transplanting rice: India.
Rice plants from this nursery bed are being gathered together for transplanting.
(*Paul Popper*)

rice plants being transplanted to the muddy padi fields from the nursery bed in which the grains or seeds are sown. The padi fields have to be flooded because lowland rice needs much water, not only in the soil but also on the growing plant. When the rice is ready for harvesting the irrigation water is drained off. The rice plants are then cut by sickle, and stacked in bundles rather like stooks of oats to dry. The grain is separated from the plant by being trodden out by oxen or water buffalo, or by being hand threshed. Rice gives a high yield per ha, but it is nearly all eaten in the countries where it is grown. Only Burma, Thailand and the Khmer Republic have a surplus for export; this they sell mainly to India, Pakistan and other hungry monsoon countries, so that little Asiatic rice enters world trade.

Tea is also an important crop in India (page 76), Sri Lanka, South China and Japan. It is useful for higher slopes, since it can withstand slight frosts and flourishes where there is 1250 mm of rain or more, provided that the land is well drained. Harvesting consists of picking off the new leaves every ten days of the wet monsoon, so much labour is needed. Tea forms an important cash crop in world markets. Many other crops are grown in monsoon lands, such as sugar cane, millet, wheat, peas, beans and other vegetables, but despite the hard work of the farmers, the vast population often go hungry. The production of food in these underdeveloped, overpopulated countries (page 157) is the greatest problem of lands with a monsoon climate.

Work to do

1. *Copy the table of Bombay's temperature and rainfall in a line across your page. Then, by means of lines, divide the months up carefully into those of the cool season, the hot season and the rainy season (page 182). Write these headings under the correct months. Give your work a title.*

2. *Draw temperature and rainfall graphs for Bombay as an example of a town with a tropical monsoon climate. Add notes to explain what the graphs show.*

3.   i) *What is meant by the Indian sub-continent?*
    ii) *Why are parts of it liable to famine?*
    iii) *Draw a map of India to show the parts liable to famine.*
    iv) *Describe how the Government is trying to prevent famine.*

4. *Make a small project file on either rice cultivation or tea cultivation.*

5. *Write an essay on 'Irrigation in India'.*

4. Deserts

As fig. 8.9 shows, more than a third of the land surface of the world is desert or semi-desert. The map also shows that most of the great deserts lie on or close to the Tropics of Cancer and Capricorn, so that daily temperatures are high. In the dry desert air there are seldom clouds, so that there is no shield against sunlight. High temperatures and very low rainfall mean that no reserve of water accumulates in the soil, and no rivers rise in them. The Nile, Indus, Tigris, Euphrates and Oxus rivers, which flow across deserts, have their source of water in other, rainy areas.

mountains and hills of desert areas suddenly fill with raging streams, carrying with them rocks, pebbles and sand. If the rain falls on broad, rocky plains some of it may seep into the ground, but much forms into very shallow lakes. The rain lasts only for a short time, however, and the water is soon evaporated in the heat. The high rate of evaporation is a great disadvantage, for not only does the water disappear but mineral salts may be left on the ground in patches, generally known as salt flats. In semi-desert areas, where more frequent rains enable large shallow lakes to survive, the water becomes very salt, so that lakes, such as the Great Salt Lake in Utah, Lake Chad in the Sahara and Lake Eyre in Australia, are little used.

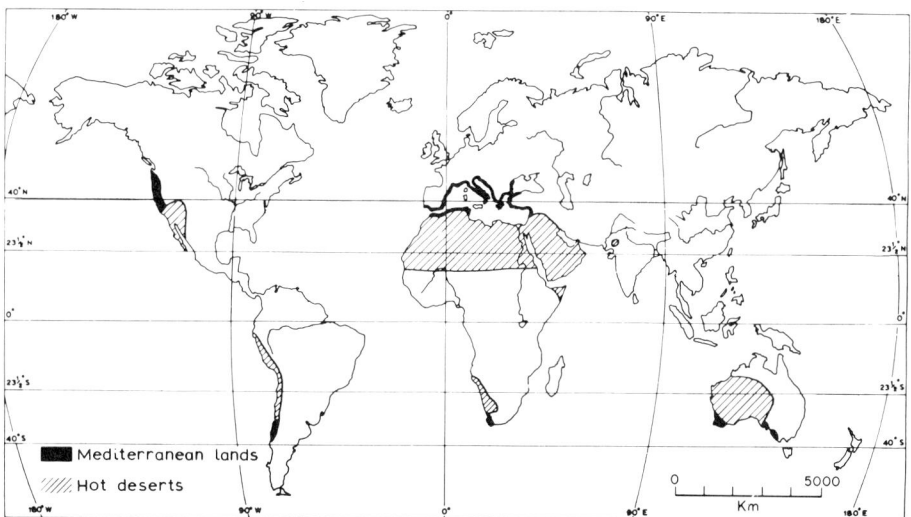

Fig. 8.9. Hot deserts and Mediterranean lands.

Mediterranean lands

Hot deserts

You will notice that we have not said that there is no rainfall in deserts or semi-deserts. Hot air rising from the land in convection currents sometimes develops clouds which build up until a thunderstorm breaks, often with torrential rain. Then gullies in the

Despite the very limited rainfall of deserts, some plants manage to survive. No plant can live without water. Desert plants have not only to overcome the water shortage but have to protect themselves against heat and tolerate a certain amount of salt.

Desert vegetation: Nevada.
Describe how the different plants are adapted to drought.
(*U.S. Information Service*)

Some of their seeds have a hard case or cover and lie safely in the soil until rain falls. As soon as moisture is available, the seed begins to grow; within a few days the plant appears, flowers and produces seeds before it dies. One type of desert grass completes its life cycle within ten days; other grasses may live for three or four weeks. Some plants having produced seed break off just above the ground, and are blown about by the wind so that their seeds are scattered. These are called tumbleweeds, and are often shown rolling across the land in television film stories set in the semi-desert areas of North America. For a short while after any rainstorm these quickly growing plants transform parts of the

desert into areas of flowers and green; they are called ephemerals.

Plants which live from one year to the next are called perennials. In the desert they include bushes, such as the acacia and tamarind. These have long roots to tap any water which may lie well below the land surface; the roots spread over a wide area, so that the bushes, too, are scattered. They may have no leaves, or have spines instead. Other perennials, such as cacti, store water in the pulpy tissues of their stems. They have thick, waxy skins so that they do not lose water by transpiration. Through shallow but widely spread roots they take in all the water they can whenever there is rain, so that they manage to survive through the long droughts. The date palm, which we commonly think of as a desert plant, is not a true example, for it can survive only where there is an adequate water supply. It will not grow in really dry areas, but only where there is an oasis or spring. It is a valuable plant, because it can survive on water which is slightly more salty than other fruit trees or crops could bear.

Lizards, snakes and spiders are common in deserts. They require little water, and can gain shade by burrowing into the ground or hiding under rocks. Wild animals are small, like the desert rat, which obtains moisture by eating roots and plant stems, and spends the hottest months of the year inactive in its burrow. The most important animal is the camel. The camel's hump is composed of fatty tissue, a reserve of fat which enables it to travel long distances without food. The camel's body heat can rise several degrees during the day without the camel sweating or panting to cool itself, so that it loses little moisture through evaporation. It can therefore live and work when the total amount of water in its body is reduced to a point at which other animals would die. After several days without water a camel can drink more than 100 litres in ten minutes, swelling almost visibly as it does so. As it has tough lips, it can eat thorny and spiny plants without harm. It has large padded feet which do not sink in sand and are hardened against rocky or stony desert. It also has thick eyelashes, which protect its eyes from sand in the air.

The surface of hot deserts contains much sand and dust, and one characteristic of desert climates which is disliked by all forms of life is their windiness. The heat gives rise to strong convection currents which result in swirling winds which whip up the dust and carry it along in clouds. These are called dust-devils; they are vortices (swirls) of hot air which occur in the daytime. More unpleasant are periodic winds, such as the khamsin, an oppressive hot wind of Egypt which blows northwards at intervals for about fifty days in March, April and May. There is little to break the force of the wind in the desert, so that travelling may be a struggle against winds carrying dust and sand. Heat is another problem. Lack of rain means the absence of cloud, and the intensely blue, clear skies allow the sun's rays to beat down all day long. The sand, rocks and plateaux reflect the sun's rays, increasing heat and aridity. Temperatures of 56·7° C have been recorded in Death Valley, California, and of 58° C in Tripolitania, North Africa. After the sun has set the clear, starlit skies allow the air to cool very rapidly, so that the nights are cold, and slight frosts may occur. This daily range of temperature is known as the diurnal range; this is very great in deserts. Because of these extremes, the Arabs of the Sahara, for example, wear heavy clothing to protect themselves from the chilly nights.

In deserts man is sometimes a nomad, driving herds of goats and sheep in search of pasture, and living in tents which are easily transported. Man can only settle in deserts if water is available. Water sometimes occurs in oases (fig. 2.4). An oasis may be a pool of water, or the water may be reached by means of wells (fig. 8.10). Where enough of such water is available man can irrigate crops of vegetables, grain and groves of date palms, and build permanent homes. Where minerals occur in deserts, such as copper at Chuquicamata in northern Chile, or iron ore at Fort Giraud

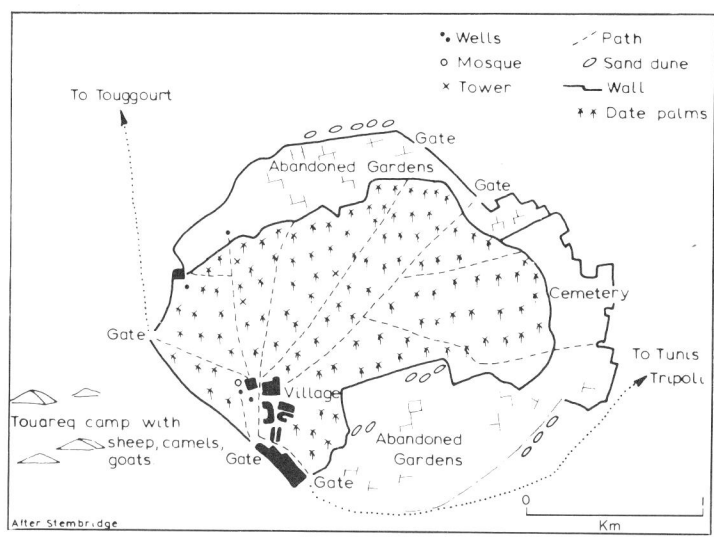

Fig. 8.10. Ghadames oasis.

Oil well in the desert: Western Australia. ▶
(*Shell*)

189

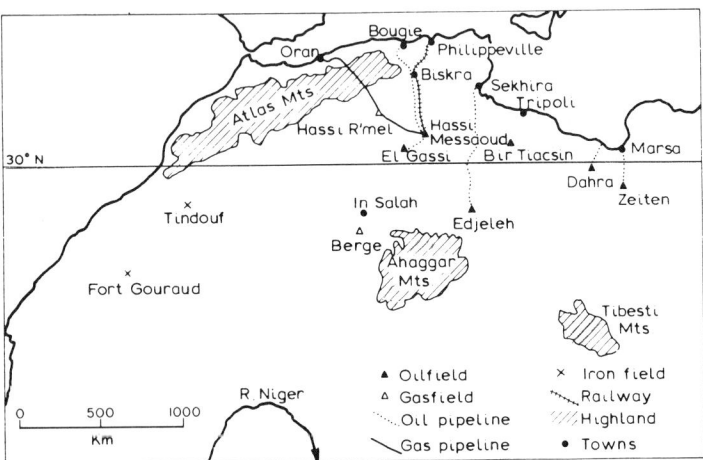

Fig. 8.11. Minerals of the Sahara.

Transport in a desert: North Africa.
See page 125
(*Shell*)

in the Sahara, it is worthwhile for man to pipe water to the mineral areas (fig. 8.11). As we saw on page 42, if oil is found near the sea coast in deserts desalination plants may provide water for the engineers and other oilfield workers.

Transport is a problem in deserts. Vehicles with caterpillar tracks are sometimes used, or lorries with heavy tyres. The wear on tracks is considerable, and no garages exist for stocking up with petrol or doing repairs in the open desert. There are few railways, and these are confined mainly to mineral bearing areas. The camel is still the most useful transport animal, but nowadays many travellers across desert use aeroplanes. Heat is another problem. The modern buildings which are constructed at mines and oil wells are nowadays air-conditioned, which makes work in them more attractive. The greatest problem is, of course, lack of water, and until man has discovered an economic way of bringing water to the great desert areas on a very large scale they will remain the scantily populated areas of the world.

Desert town: Kuwait.
Make notes on (i) the old and new kinds of transport, (ii) the clothing of the people, (iii) the shape and style of the houses.
(*Camera Press*)

Work to do

1. *Write a paragraph describing the climate of the Sahara Desert.*
2. *Describe the ways in which plants manage to survive in deserts.*
3. *List the ways in which life is changing in deserts today.*
4. *The following are the temperature figures for Aswan (24° N., 33° E):*

	Jan.	Feb.	March	Apr.	May	June
Temperature, °C	15	17·2	21·1	25·6	29·4	32·2
Rainfall, mm			Practically nil			

	July	Aug.	Sept.	Oct.	Nov.	Dec.
Temperature, °C	32·8	32·2	31·1	27·8	22·2	16·7
Rainfall, mm			Practically nil			

   i) *Which are the cool months in Aswan?*
  ii) *Rainfall is practically nil. Under what conditions might some rain occur?*
 iii) *Find Aswan on an atlas map. From what source does the city obtain water?*
 iv) *How do farmers living on the outskirts of Aswan manage to grow crops?*

5. *Study the map of the oasis (fig. 8.10) and answer the questions:*
   i) *What surrounds the oasis? Why do you think it was built?*
  ii) *How many entrances are there to the oasis?*
 iii) *Where do the people of the oasis get water?*
 iv) *What is the chief crop of the oasis?*
  v) *What does the word garden suggest about the type of cultivation here?*
 vi) *Why have some gardens been abandoned? The word dunes gives you a clue.*
 vii) *Describe the position of the village.*
viii) *Why do you think towers were built? They are not used now.*
 ix) *What is the religion of the people? How do you know?*
  x) *Who are the Touaregs? Why are they camped outside the oasis? For what reason have they come to the oasis?*

5. Mediterranean lands

'The roads went like a white switch-back between little valleys in steep cliffs of rock that seemed cracked and roasted in the great glare of the sun. In the valleys were also many little orchards of lemon and orange and a few of peach and almond. Old walls of stone, crumbling away, disappeared into entanglements of bushes and vine. . . . In the heat of the morning very little moved; a woman drawing water from a standpipe and carrying it away in a rose-brown pitcher on her head, a few scrawny sheep panting in the shade of the trees, a goat in the white dust of the road, an incredibly small donkey bearing a woman up a path, scattering a red hen and her brood of yellow chicks. . . . The tranquillity of it all spread through him as he sat under the black shade of an olive tree of enormous spread and great age. . . .'

This description is of Greece, but it could have been written of parts of Spain, southern France, the coastlands of Italy, Turkey and other countries, and islands bordering the Mediterranean Sea. If you read it again you will discover phrases which suggest that the season is summer, that the land is agricultural and that water is not laid on in all the houses as one expects in our country. Indeed, water in these Mediterranean countries creates a problem. Let us look at another description which will help to explain it.

'From both sides of the valley little streams slipped out of the hill canyons and fell into the bed of the Salinas River. In the winter of wet years the streams ran full, and they swelled the river until sometimes it raged and boiled, bank-full, and then it was a destroyer. The river tore the edge of the farmlands and washed whole acres down; it toppled barns and houses into itself, to go floating and bobbing away to the sea. . . . Then when the late spring came, the river drew in from its edges and the sandbanks appeared. And in the summer the river didn't run at all above ground. Some pools would be left in the deep swirl places under a high bank. . . . The Salinas was only a part-time river. The

summer sun drove it underground. . . . And from then on until the next rains the earth dried and the streams stopped. Cracks appeared on the level ground. The Salinas River sank under its sand.'

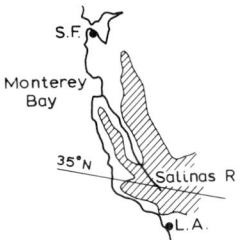

Fig. 8.12. The Salinas River.

The Salinas River is in California, and flows out into the Pacific Ocean at Monterey Bay (fig. 8.12). The climate of California, and of other lands shown on the map (fig. 8.9), is very similar to that of the Mediterranean coastlands; this type of climate is called 'Mediterranean'. You will have realised why water is a problem. Rain falls only during the winter months, while the summer months are practically rainless. In figures the rainfall and temperature look as follows:

Jerusalem	Jan.	Feb.	March	Apr.	May	June
Temperature, °C	6·7	8·9	10·6	15	18·9	21·1
Rainfall, mm	157	117	89	38	8	—

	July	Aug.	Sept.	Oct.	Nov.	Dec.
Temperature, °C	22·8	22·8	21·7	19·4	13·3	9·4
Rainfall, mm	—	—	—	10	64	145

The winter months are the rainy months, but temperatures then are very mild.

In the summer months, when the skies are cloudless and temperatures are high in the long, sunny days, only long-rooted trees like the olive will grow without irrigation. Yet just when the rivers might be useful for this purpose, some of them dry up, as the Salinas River used to, disappearing save for a few pools. Fortunately, along the northern shores of the Mediterranean Sea, in California and in central Chile, the lowlands are backed by mountainous country. Here the ground rises so high that deep snow falls during the winter months and the mountain summits are permanently capped with snow. When this melts in spring and summer it increases the flow of water to the rivers, and these can be used for irrigation. Often they are dammed so that their water is held back in reservoirs, from which it is permitted to flow gradually during the summer. In this way floods like the one described are nowadays generally avoided.

As so little rain falls in the summer months, the natural vegetation of the Mediterranean lands is adapted to withstand summer drought. Many of the small evergreen bushes, such as lavender, thyme and rosemary, are highly scented. They grow their new leaves and flower during the winter rains, making the countryside very attractive. In summer they turn brown or dull grey-green because of the dry heat and excessive evaporation. There are evergreen trees like the Corsican pine or the cork oak, from the bark of which cork is stripped. There is not a great deal of grass, because the summer heat withers it. The summer heat also affects the lives of the people and influences building.

Since temperatures in the Mediterranean coastlands are well above freezing all the year round, it is possible to grow crops at all times if sufficient water is available. The winter rains from November to March enable the farmers to grow grain, mainly wheat and barley, planted in autumn and harvested in spring. These rains also make possible the cultivation of potatoes, carrots, beans and onions, together with salad crops, such as tomatoes, radishes, cucumber and lettuce. These 'early' vegetables and market-garden crops often appear in our shops in early spring long before our own crops are ready for harvesting.

193

In summer the farmer can only grow crops if he can irrigate them. Maize, which requires high temperatures and much sun for ripening, is grown. The picture on page 50 shows you how water is spread around the fields. Peaches, apricots and walnuts also ripen during the summer. In some areas there are groves of oranges and lemons. These grow on trees which are very sensitive to frost (fig. 1.4). The fruit usually ripens in September and October; picking and packing is a busy time. In Spain they not only grow sweet oranges but bitter oranges known as Seville oranges for use in making marmalade.

Many of the farmers own vineyards. The vine from which the grape is harvested is another plant which flourishes in Mediterranean lands. It is usually pruned in March to keep it from growing very high and to encourage new growth. During the summer the vines are sprayed to prevent disease, and the ground around the vines is hoed to keep it free from weeds. In September the farmer and his family are very busy gathering the grapes. Sometimes they are sold as fruit. In Greece and California different varieties are dried to make currants, sultanas and raisins. More often they are squeezed in a wine press to make wine. There

Bottling sherry at Jerez: Spain.
(*J. Allan Cash*)

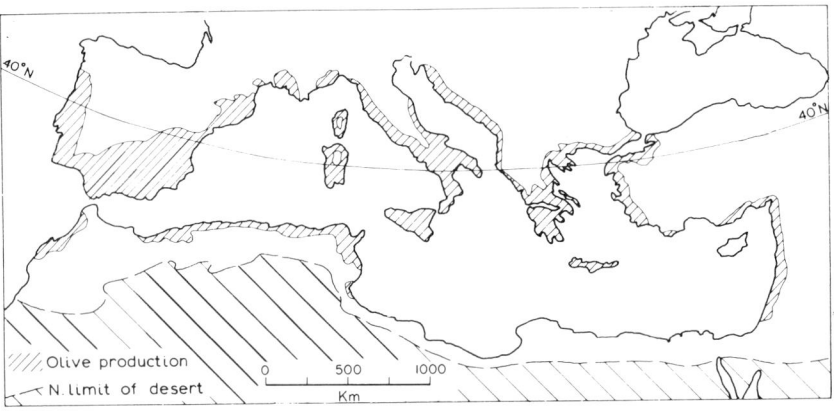

Fig. 8.13. Olive growing areas in the Mediterranean.

are many famous wines, but perhaps the best known is sherry, which is made from grapes grown around Jerez, in Spain.

One of the most valuable trees is the olive, which can thrive on little water. It has a thick, gnarled bark and small leaves well adapted to a dry climate. It does not mind dry, stony soils, and is often grown in groves on hillsides which might otherwise be useless. The fruit of the olive tree is like a small, green plum. It can be eaten as it is, but is often pressed to squeeze out the valuable oil it contains. This is olive oil, which has many uses, but is mainly a cooking oil, used instead of butter, margarine or lard. The olive is harvested in December and January. If the olives are due to be crushed they are beaten off the branches with the aid of long sticks. They fall on to cloths spread under the trees, from which they can easily be picked up. Where olives can grow is often taken as the boundary of the true Mediterranean climate (fig. 8.13).

Farming in the narrow coastal lands round the Mediterranean Sea is usually on a small scale. In Central California early vegetables and salads are grown on a large scale in enormous fields. The olive is less common, but there are thousands of acres of

Large scale farming: vineyards in Victoria, Australia.
(Australian Information Bureau)

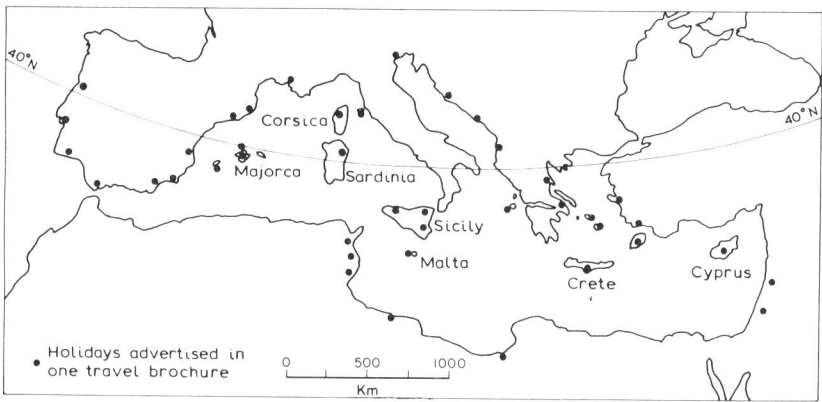

Fig. 8.14. Holiday resorts in the Mediterranean.

orange groves. In order to prevent stealing, policemen on motor cycles constantly patrol the fruit areas at harvest-time.

The countries with a Mediterranean climate are warm, showery, green and fresh in winter; attractive with fruit blossom in spring; very hot, sunny and dry in summer; and pleasantly warm for the harvests of autumn. No wonder that they are holiday areas all the year round (fig. 8.14). For the peasant farmers who live there, however, work on the farms is hard. In Spain, Italy and Greece many of the farms are too small to support the whole family, so the sons emigrate. They find work in countries like Australia, Brazil and Argentine. They always plan to return home when they are old, to spend peaceful days in the sunshine of their Mediterranean lands.

Orange grove: California. ▶
See also page 51 and fig. 1.4.
(*U.S. Information Service*)

195

Work to do

1. *Write a paragraph describing how farming in California differs from that in European lands, using the pictures to help you.*

2. *Draw temperature and rainfall graphs of Jerusalem (page 192) as an example of a town with a Mediterranean climate. Add notes to explain what the graphs show.*

3.

Activity	Spring	Summer	Autumn	Winter
What the weather is like				
What the vegetation is like				
What crops are harvested				

Copy this table, if possible making it larger. Then complete it for Mediterranean lands.

4. 'The formal beds beneath the peach trees were rich with thyme and lavender and purple rosemary, while the feet of the pear and apple trees espaliered on the surrounding walls stood deep in a silver drift of sage. A row of apricot trees lent support to a disciplined riot of vines; below it, in careful ranks, fading stems were weighted with the fabulous red of tomatoes. There was even a pair of orange trees. . . . Over all hung the scent of the near pinewoods.'

   i) *What do you think the paragraph is describing?*

   ii) *In what climatic region is the area located? How do you know?*

   iii) *What two trees are not typical of this area, that is will grow in other regions?*

   iv) *Which is the typical tree named?*

   v) *What are thyme, rosemary and sage used for?*

   vi) *What particular aspect of climate do peaches, pears, apples and apricots need which makes them flourish in this area?*

5. *Why is water a problem in Mediterranean lands? How is the problem solved?*

6. Temperate grasslands

The map (fig. 8.15) shows you the five areas we are going to discuss in this chapter. They are the Steppes of the U.S.S.R., the Prairies of North America, the Pampas of South America, the High Veld of South Africa and the Murray–Darling Basin of Australia.

Here is a description of the natural grassland in spring. 'Only when the life-giving sunshine is accompanied by the soft south wind, at the earliest in the beginning of April, usually about the middle of the month, does the snow disappear quickly. Even before the last snow wreaths have vanished, before the ice-blocks have melted on the lake, the bulbous plants and others put forth their leaves and raise their flower-stalks to the sun. Among the yellow grass and the grey stems the first green shimmers . . .

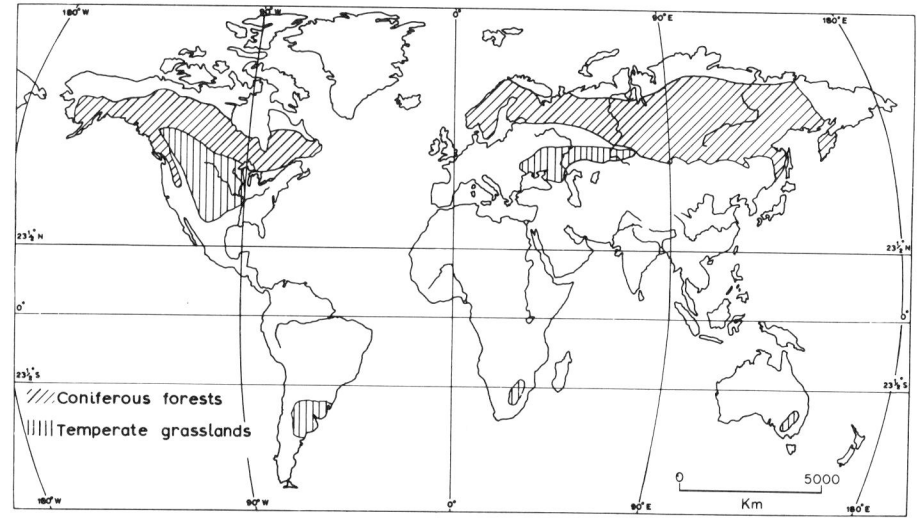

Fig. 8.15. Temperate grasslands and coniferous forests.

The Steppes of the U.S.S.R. extend from the Ukraine in the west across Siberia almost to Lake Baikal. They are natural grasslands. It is only in eastern Siberia that these grasslands remain in their natural state, for most of them have been ploughed for cultivation or for sowing with new grass seed to form improved pasture land.

boundless tracts are resplendent with tulips, yellow, dark red, white, white and red. They rise singly or in twos and threes, but they are spread over the whole steppeland, and flower at the same time, so that one sees them everywhere. Immediately after the tulips come the lilies . . . . After a few weeks the steppeland lies like a gay carpet in which all tints show distinctly.' The heat of the summer soon withers the flowers and turns the grass yellow. After a short autumn 'a single night's frost covers the lakes with thin ice. Gentle north-west winds sweep dark clouds across the sky, and the snow drizzles down in small flakes. . . . The wind changes . . . a thin cloud sweeps over the white ground—it is formed of whirling snow; the wind becomes a tempest; the cloud rises up to heaven; and the buran (a snow hurricane) ranges across the steppe.' The long winter has begun.

The wind can rage across the Steppes because they are open, flat plains. Nomadic tribes are very rare, since the Steppes are being cultivated farther and farther east, but some Khirgiz herdsmen still move with their cattle. They put up tents of felt and hide stretched over a lattice framework. These are called yurts; they are round, because this enables them to withstand strong winds better. The lattice framework is kept carefully because wood is scarce, since there are few trees. Over the vast areas where the Steppes are cultivated man often plants trees round his house to form a windbreak. The houses are usually clustered together because the people work together on state farms and collective farms.

A state farm is owned by the Government and is managed by workers who are paid wages. A collective farm is owned by the farmers who work on it. After harvest some of its produce is collected by the Government. some is sold to provide recreational facilities for the farmers and their families, the rest is shared among them in proportion to the number of hours they have worked. The farmers also have a small plot of land next to their houses on which they can grow vegetables for the family. Both types of farms are large, often having several hundred workers. They are completely mechanised, that is ploughing, seeding, hoeing and harvesting is done by machinery.

The U.S.S.R. is the world's greatest wheat producer, and some sixty per cent of this grain is grown in the Steppes. The Ukraine is a particularly important area as the soil over large areas is 'black earth'. This is a very fertile black soil rich in humus or decayed

Driving cattle: Nebraska.
See also picture on page 179.
(*U.S. Information Service*)

vegetable matter, in which plants flourish. Other crops such as barley and sugar beet are grown besides wheat.

The temperate grasslands or Prairies of North America were originally the home of the bison and the Red Indian. As fig. 8.16 shows you, large areas particularly in the United States are grazing lands for cattle. In South Dakota, Nebraska and Kansas most of the natural grass has been replaced by more nourishing, improved varieties. Cattle ranches are generally very large; these are the home of the cowboy. Nowadays he seldom carries a gun and often owns a car. On some ranches he looks for missing cattle from a helicopter. His work still consists of rounding up cattle for counting, branding, dipping and selection for market. Cattle are

Winter scene: Prairies.
Why is this winter feeding necessary?
(*U.S. Information Service*)

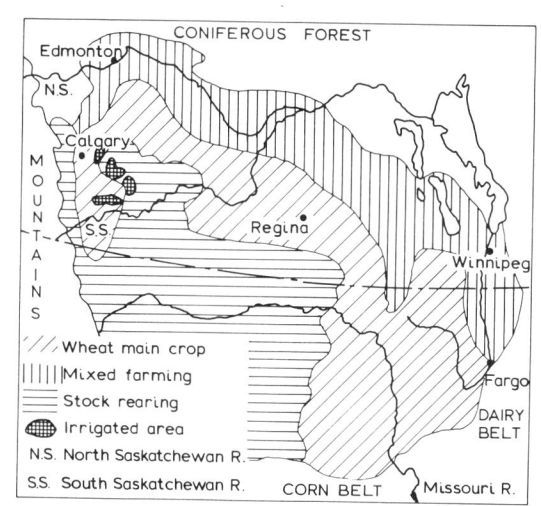

Fig. 8.16. Farming belts in the Prairies.

sent by train for sale in the towns of Kansas City, St. Louis and Chicago, not driven on the hoof for long distances as they were in the nineteenth century. Many of them are fattened up on the farms of the Corn (Maize) Belt before slaughter.

Those parts of the Prairies which form the great spring wheat belt lie in southern Alberta, Saskatchewan and Manitoba in Canada, and in North Dakota and the north-east of South Dakota in the United States (fig. 8.16). In our country wheat is sown in autumn of one year and harvested in late summer of the next year; this is 'winter' wheat. In the Prairies, as in the Steppes of the U.S.S.R., wheat is sown in spring and harvested at the end of summer in the same year. This 'spring' wheat is of varieties which

will grow and ripen within 110 days. As the table below shows, winter temperatures are below freezing for five months; seeds planted in autumn would not live during winter. Add up the rainfall figures; you will find the total is less than that of London (610 mm). Some of this falls as snow during the winter. When spring comes some of the snow melt soaks into the ground. In drier parts of the wheat belt some of the fields are left fallow for a year (page 70); in this way the snow melt of two winters can be left in the ground to produce a better crop in the second summer. Fallow fields are also seen in the Russian Steppes.

Winnipeg	*Jan.*	*Feb.*	*March*	*Apr.*	*May*	*June*
Temperature, °C	−20	−17·8	−9·4	3·3	11·1	16·7
Rainfall, mm	23	18	30	36	51	79

	July	*Aug.*	*Sept.*	*Oct.*	*Nov.*	*Dec.*
Temperature, °C	18·9	17·8	12·2	5	−6·1	−14·4
Rainfall, mm	79	56	56	36	28	23

Combine harvesters in a wheat field: North Dakota.
(U.S. Information Service)

The map (fig. 8.17) shows a wheat farm in southern Saskatchewan. You see that the farmer grows barley, peas and flax besides wheat. Barley is sold for cattle food, peas for canning; flax seed is sold to factories, where it is crushed to produce oil called linseed oil, which is used in the production of paint, varnish, linoleum and patent leather. The farmer grows these plants so that he need not rely solely on a good wheat harvest. The wheat from this farm and hundreds of similar farms is collected in elevators, tall storage buildings seen in the picture. These have been built at intervals along the railways of the wheat belt. The map (fig. 8.18) shows you Canadian grain export routes. The wheat of the United States is collected in towns like Minneapolis to be distributed over the United States and sent to ports like New York for export.

The areas of temperate grassland in the southern hemisphere lie

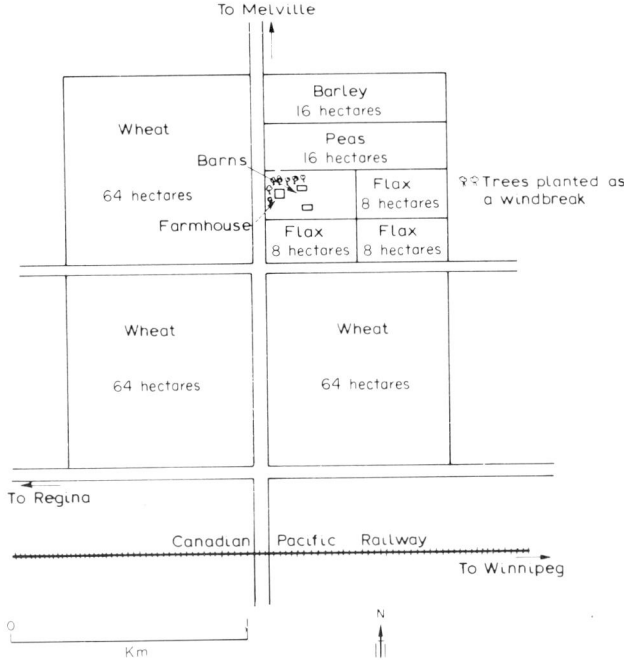

Fig. 8.17. A Prairie farm.

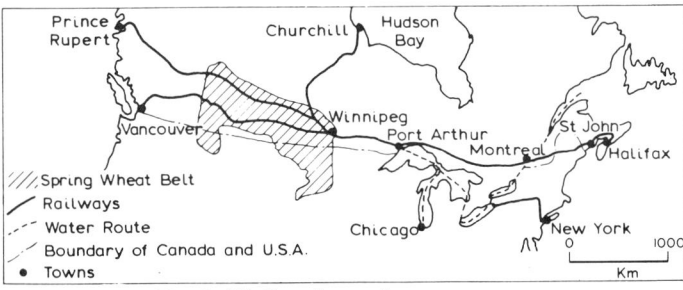

Fig. 8.18. Canadian grain export routes.

Loading grain from an elevator: Great Lakes.
This is a specialised bulk carrying vessel which plies on the Great
Lakes. The grain in bulk can be handled as a liquid.
(*U.S. Information Service*)

nearer to the Equator than the Steppes and Prairies, so they
experience higher temperatures all the year round. The High Veld
differs from all others because it is plateau land at an elevation of
600–1800 metres (fig. 8.19). Maize is the grain crop over large
areas, but as rainfall is unreliable, yields are low. The drier west
supports sheep; the grass is better in the east, so cattle are kept.

```
                               Maize and
            Semi-desert   Sheep   cattle
Atlantic                                    Indian
Ocean                                       Ocean
```

The Pampas is the great farming area not only of Argentina but of
the whole of South America. Maize, wheat and flax are grown
over vast areas. The Pampas also has great cattle ranches called
estancias. A single estancia may have over 4000 cattle. The
original Pampas grass, which we sometimes see in gardens, has
been replaced by imported European grasses and by alfalfa, a
deep-rooted plant belonging to the clover family. As few estancias
have rivers or streams, water is drawn from wells by means of
wind pumps. The lack of surface streams is due to the extreme
flatness of the Pampas, for over large areas after heavy downpours
the rain remains as a sheet of water until it finally sinks into the

soil. Wind pumps are also a common feature in the Murray–Darling Basin of Australia, where there are many sheep farms. This third area is not like the open grasslands of the Pampas, but is a parkland area on the east, where many of the trees have been removed. Here the farmers grow not only wheat, oats and barley over half their farm but also let sheep graze on the uncultivated part. In the west the land becomes poor scrub which can only support sheep. Where water is available the land is irrigated and farming is different (fig. 2.13).

It is clear that one of the problems of temperate grasslands is low and sometimes unreliable rainfall. Attempts are made to overcome this disadvantage by leaving land fallow. Farmers also sow their grain seeds thinly over very large areas. The large farms are worked by relatively few men with many machines, and though this extensive farming gives low yields of grain per ha, the temperate grasslands are the granaries of the world. The dangers of depending on a single crop like wheat are being minimised by the planting of other crops, and mixed farming is becoming more common. The long cold winters of the Prairies, for example, when all farming of land is at a standstill, prove too severe and monotonous for some farm workers, many of whom are leaving the land to find work elsewhere. Finding labour is thus another problem in some areas. Where stock are kept, the farmer has work all the year round, but he may be faced with the problem of finding winter feed. Apart from climatic problems, the temperate grasslands are generally good agricultural regions, upon whose produce much of the rest of the world depends.

Work to do

1. *Write a description of the climate of either the Prairies or the Steppes.*

2. *Study the map of the Prairie farm* (fig. 8.17) *and answer the questions:*

   i) *What is the size of the farm?*
   ii) *What proportion of the farm is given to wheat growing?*
   iii) *What area is given to the house, barns and their grounds?*
   iv) *Why are trees planted round the house and barn?*
   v) *In which town will the crops be sold?*
   vi) *Why are beans, peas and flax grown?* (*see text*).

3. *Write an account of a year's work of a Prairie farmer.*

4. *Using your atlas, draw a map of the High Veld. This extends west of the Drakensberg Mountains from Johannesburg over the southern Transvaal and the Orange Free State to the Orange River. Name these features. Add a note to say how this area differs from other temperate grasslands.*

5. *Why has most of the natural grassland of the areas discussed in the chapter disappeared? What has man put in its place?*

7. Coniferous forests

A journey through the coniferous forest or taiga gives an impression of great monotony. 'For hundreds of kilometres the dense cover of evergreen conifers casts an unbroken shadow on the ground beneath. In consequence, undergrowth is scanty or non-existent, the ground being covered with a litter of needle-shaped leaves and dead, decaying wood. The sombre, shadowy tones of trunks, branches and foliage are rarely relieved by the brighter hues of grass and flowers.'

Coniferous means bearing cones; the cones contain the seeds of the trees. There are many kinds of coniferous trees. From Norway to the Ural Mountains the pine is most common; in the forests of Asia are stands of larch and spruce as well. In Quebec and Ontario, on the southern part of the Laurentian Shield, white and black spruce and balsam fir are abundant, while the forest which stretches north through the Prairie Provinces and westwards to British Columbia consists mainly of lodgepole pine. All these lands endure long, harshly cold winters, when the soil is frozen. The needle shape of the leaves reduces their area, and so reduces transpiration or loss of water from the leaves, a necessity when water cannot be obtained from the frozen ground. Conifers are evergreen trees; they do not lose all their leaves at any season. They do not have to wait to grow new leaves before they can profit from the short spring and summer warmth.

These northern coniferous forests are of great value to man (fig. 8.15). In them live fur-bearing animals such as the marten, ermine, fox, bear and beaver. It was these animals who first attracted man to the forests, but fur-trapping is now a declining if exciting occupation. The fur-trappers work from log huts built from the forest timber, to which they travel by canoe or sledge in late autumn, before the rivers freeze. It is in this season that the animals' fur becomes thick and glossy in preparation for winter. The trapper usually hides forty or fifty traps in the snow each day,

on a circular trail which leads him back to his hut. At intervals he visits them to remove his catch and reset the traps. During the long, lonely winter he stores the furs which have been removed and dried out in his hut, then in spring he takes them by sledge or canoe to a trading post. Here, in exchange, he buys all that he will need for the next winter's trapping, leaves it in store and either has a holiday during the summer or goes to find some other work elsewhere.

More valuable than the furs is the timber of these forests. In Canada the forests provided wood for the fuel, homes, barns and ships of the early settlers; nowadays forest industries, which include logging or lumbering, sawmilling and the manufacture of pulp and paper, provide about one-fourth of Canada's exports, bought largely by the United States. In Sweden timber mills

Transporting timber.
(*U.S. Information Service*)

employ as many people as those in the iron mines, machine and other metal factories, and the United Kingdom buys much Swedish timber. In the U.S.S.R. more than half of the yearly output of timber is used as fuel, and since paper consumption is estimated at being 15 kg per person compared with 198 kg in the United States, there is less paper making.

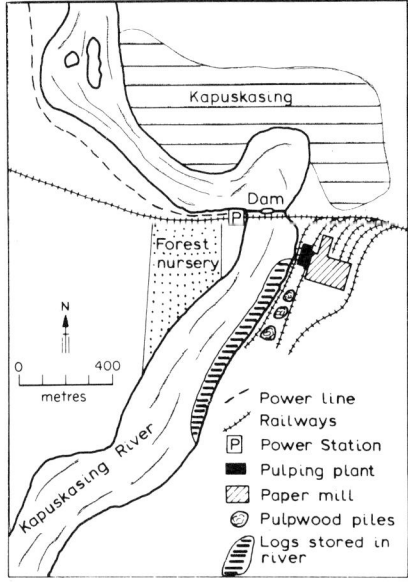

Fig. 8.20. Kapuskasing paper mill: North Ontario.

Let us look at one of the largest paper mills in the world, located at Kapuskasing in Northern Ontario (fig. 8.20). It can produce 900 tonnes of newsprint a day. The biggest problem is ensuring a constant supply of wood. This is cut from 18 000 square km of forest leased from the Government. In summer 1000 men are employed in the forests, selecting trees for felling,

clearing tracks and planting out four-year-old seedlings taken from the company's nursery to replace felled trees. In winter twice as many men are employed. They fell the trees, remove branches and load the timber on to trucks or trains for transport to the mills. Some of the timber is stacked up by the riverside to await the spring thaw, when it is floated downstream to the mill. Before the wood can be used it is sawn into metre logs, and the bark is removed by rotating these in big drums, where they are sprayed with water under pressure. Large quantities of steam are then used to 'cook' the pulpwood, and to dry the pulp when it is pressed through rollers to become paper. It has been estimated that 250 tonnes of water are used in making 1 tonne of pulp, so that a waterside location is necessary for any pulp and paper mill. Every mill also requires a great deal of electricity to produce steam, heat and energy for the various manufacturing processes, and a waterside location is often selected because it also enables the production of cheap hydro-electricity.

Transportation is also important in lumbering, because timber is bulky and heavy in proportion to its value. Moreover, lumber comes from areas where the population is scanty and there are relatively few roads. Winter snow facilitates transport, for sledges pulled by caterpillar tractors can slide the timber over the smooth snow surface to the nearest river. In Sweden, for example, where there are many short rivers flowing east to the coast, a haul of only 3 or 4 km brings most timber to a river. The logs are piled on the ice, and the spring thaw carries the timber to the mills near the seashore. Drivers with long poles or pevees prevent the floating timber from jamming at river bends or over rapids. In calm water the logs are collected into rafts and pulled by tugs. Much of Sweden's timber is made into kitchen furniture, doors and window-frames.

Man's use of the coniferous forest is not without problems. The areas of forest nearest to habitation have been cut ruthlessly so that they have been destroyed. Nowadays it is realised that if future generations are to have any timber, trees which are cut

Forest fire.
Write a story around this picture.
(*U.S. Information Service*)

down must be replaced. The replanting of trees is called re-afforestation. Growing young trees from seeds takes care and skill, and replanting is expensive. Another problem is the prevention of fire. In the summer these forests may become very dry, and since they are often used as holiday areas by tourists, a picnic fire not properly put out or a cigarette end carelessly thrown away can start a fire. Such fires spread rapidly, destroy hundreds or even thousands of hectares of valuable trees and are hard to put out. All coniferous forests cut for timber have fireguards, who keep watch from control towers high above the trees. At the first sign of smoke they radio information to fire-fighting teams, which may include pilots whose aircraft carry thousands of litres of water in tanks. The water is released round the edges of the fire to prevent it spreading. Many forests have wide paths cut through them so that flames cannot leap across the gaps between the trees. There is no more depressing sight than an area of once green and splendid trees reduced to charred, blackened stumps, devoid of all animal and bird life, ugly and desolate.

Work to do

1. *Study the picture on page 203 and answer the questions.*
   i) *What is a lumberjack?*
   ii) *Describe the scene in the picture and say what is happening.*
   iii) *How does the winter climate help lumbering?*
   iv) *For what are the logs used?*

2. *Write a paragraph to explain what use rivers are to lumbermen.*

3. i) *Name six fur-bearing animals which live in the coniferous forests.*
   ii) *During what season are they trapped? Why?*
   iii) *What does the trapper do with the furs he collects?*
   iv) *What does the trapper do in summer?*
   v) *Many fur-bearing animals are nowadays bred on farms. Suggest reasons for this.*

4. i) *Name three items necessary to establish a paper mill.*
   ii) *Why is a waterside location necessary for a paper mill?*
   iii) *Briefly describe how paper is made.*
   iv) *Name as many different types of paper as you can.*

5. *The following are the temperature and rainfall figures of Helsinki (Finland):*

	Jan.	Feb.	March	Apr.	May	June
Temperature, °C	−6·1	−6·7	−3·9	1·1	7·8	13·9
Rainfall, mm	46	36	36	36	46	46

	July	Aug.	Sept.	Oct.	Nov.	Dec.
Temperature, °C	16·7	15·6	11·1	5·6	0·0	−3·9
Rainfall, mm	56	74	63	66	63	61

   i) *In which months is 'rain' likely to fall as snow?*
   ii) *What is the total precipitation in millimetres? (Precipitation is rain and snow.)*
   iii) *How do Helsinki's summer temperatures compare with London's?*
   iv) *Draw graphs to represent the given figures.*

8. Arctic lands

The map (fig. 8.21) shows the main Arctic lands of Europe and Asia. If you study it you will see that the tundra areas shown lie mainly north of the Arctic Circle. The coastal lands are crossed by the 10° C July isotherm. This means that these lands have an average July temperature which is seldom higher than those which we ourselves experience in March. This summer heat is only sufficient to thaw out the top soil, and it is in this shallow layer of some 50 cm that the tundra vegetation has its roots. Lichens,

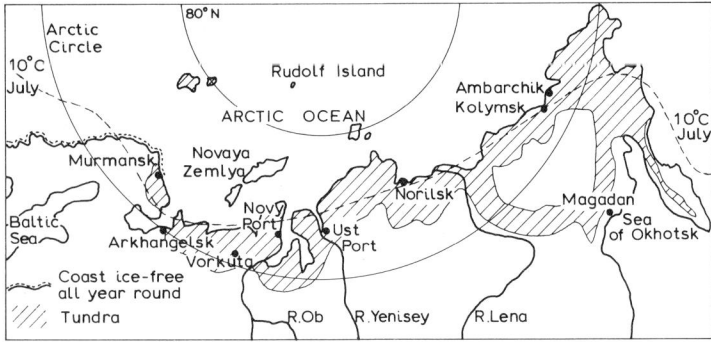

Fig. 8.21. Arctic lands in Europe and Asia.

mosses, clumps of buttercups, bilberries, tufted grasses and occasionally dwarf willows are among the small plants which brighten the brief, cool, cloudy days of the short summer. These plants begin to appear as the winter snows melt. Winter is a long, bitterly cold season, when icy air from the Pole blankets the land. Snow begins to cover the ground at the beginning of October, and remains for some 250 days. It lies only some 75 cm deep, but the intense cold, with January temperatures of −20° C, freezes the ground permanently to a depth of 1 or 2 metres. This frozen

earth does not permit the growth of long roots, and coupled with the constant fierce winds which prevent the survival of tall plants, contributes to the lack of trees in the area.

Although these lands are bordered by the sea, you will observe from the map that only in the extreme west are the shores always ice-free. These are the shores warmed by the North Atlantic Drift before it is finally lost in the cold Arctic waters. The main part of the Arctic Ocean is covered with permanent pack-ice which prevents navigation at all times (fig. 8.22). The coastal waters are free from ice for two months in late summer, but powerful ice-breakers can force their way through at other times if necessary. The map shows you that only one Russian port is open to ships all

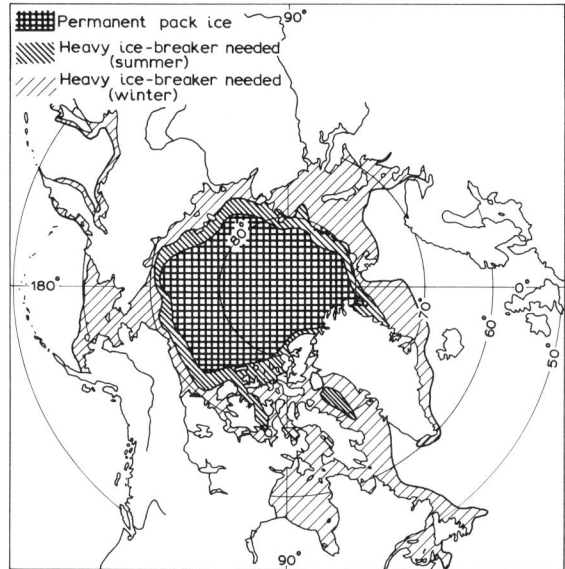

Fig. 8.22. Ice in the Arctic.

Ice-breaker in the Arctic.
(*Aerofilms*)

the year round. Indeed, transport of any kind is very difficult in this area. Constant cloud cover, fog, strong winds and the formation of ice hinder aeroplanes; navigable rivers such as the Ob, Yenesei and Lena are frozen for up to seven months of the year; and there are but few miles of road or rail. The common form of transport is by dog or reindeer sledge.

Despite difficulties of climate and transport, the Arctic lands are not completely uninhabited. Birds, polar bears, seals, walrus, fish and animals like the Arctic fox and hare enable scattered, partly nomadic tribes to live by hunting and collecting eggs, eider down and berries. Some herd reindeer, for which the tundra provides summer pasture. In recent years, however, these lands have become important for other reasons. The need for weather information has led to the establishment of many permanent Soviet weather stations in the area, including that on Rudolf Island, the

most northerly weather station in the world. Equally important, the discovery of minerals, such as lead, silver, zinc, iron and coal, is encouraging miners to endure the harsh climate. There is now commercial fishing round the mouths of the Ob and Yenesei Rivers, the fish being processed at Ust Port and Novy Port. The Russians are trying to improve diets by developing agriculture, and in some settlements limited amounts of vegetables are produced in glasshouses with heated pipes. The settlements themselves may have heated pipes under their wooden pavements and their log huts also centrally heated.

Although it is possible to make living conditions in the Arctic lands endurable by means of specially adapted clothing, the artificial heating of whole towns and the import of food, costs are high. There are schemes for producing hydro-electricity from the Lena and Yenesei Rivers, which may lower power costs. Producing fresh vegetables by artificial heat is also costly, and is thus likely to be confined to small areas only. If these lands are to be developed the improvement of transport is essential, for food, equipment and machinery must be brought into the area.

The Russians have been concerned with improving transport since the last world war. Many airfields were then established, and now there is an important air route between Vorkuta and Norilsk, as well as other local routes. The most spectacular of Soviet transport developments, however, is the opening up of the Northern Sea Route from Ambarchik to Murmansk. By means of ice-breakers, scouting aircraft and the large number of weather stations, the U.S.S.R. now provides ice forecasts, and ships proceed in convoys whenever weather conditions permit. They carry food and supplies to mining centres such as Norilsk and Kolymsk, bringing back minerals and furs. The western part of the route, between Murmansk and Ust Port, is most used; ships are fuelled by coal from Vorkuta.

Land routes in the Soviet north are still few. Murmansk and Arkhangelsk have long been linked by rail to southern Russia; since 1943 the railway has been extended to Vorkuta, and on to the west bank of the Ob. The more important all-weather road, made possible by permanently frozen ground and only light winter snowfall, runs from Magadan to Kolymsk.

The Arctic lands of North America are very similar to those we have been describing (fig. 8.23). Canada and the United States have co-operated to establish a line of radar listening posts across their northern shores. These posts are known as the Distant Early Warning (D.E.W.) line, and have been built to detect the approach of aeroplanes flying across the Polar regions from Europe or Asia. There are also Ballistic Missile Early Warning Systems Stations. There are various mining settlements. Gold is mined at Yellowknife, uranium at Port Radium, copper at Coppermine and at Norman Wells is a rich oilfield. Experiments are in progress to find out if it is possible to keep caribou, or wild reindeer, on a large scale. Many of the herds graze round Aklavik and on lands east of the delta of the Mackenzie River. Northern Alaska is developing more quickly than elsewhere, largely because the building of the Alaska Highway from Edmonton in Alberta to Fairbanks brought many new settlers to the north.

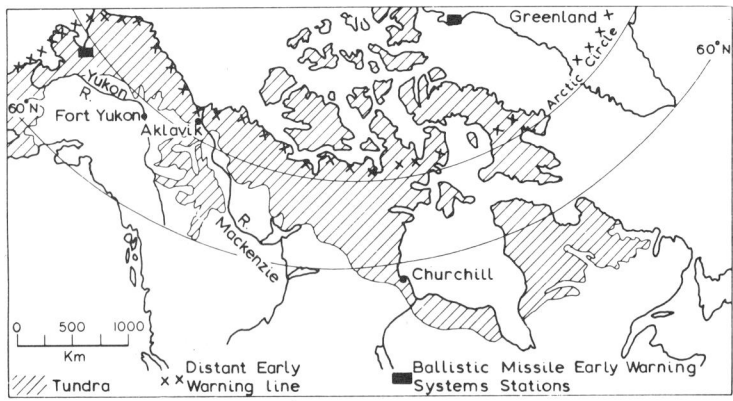

Fig. 8.23. Arctic lands in North America.

Hut interior: Alaska.
List the features in this picture which suggest cold weather outside.
(*U.S. Information Service*)

Iceberg off the coast of Greenland.
The permanent ice sheets slowly move down to the sea and float away,
as shown in this picture, to form icebergs.
(*U.S. Information Service*)

Aeroplanes now supply even remote settlements with meat, butter, milk and vegetables, and with fuel oil and equipment. In winter tractor trains, which carry 70 tonnes of freight and up to twenty passengers, and have powerful headlamps to overcome the Arctic darkness, travel over the frozen land.

It does not seem likely that the Arctic lands will ever support a dense population. Nevertheless, there are already many more people living there than was once thought possible. The increasing demand for minerals, of which there are rich pockets in the tundra lands, is probably the most powerful influence in persuading people to work there. Although Arctic summers are short, the many hours of daylight may yet allow the growth of specially developed hardy cereals or crops on a worthwhile scale. Apart from clouds of mosquitoes which appear in some parts when the snow thaws, these lands are singularly free from pests and germs. It is the long, dark, dreary winter which is man's enemy.

The Antarctic is even more inhospitable. It consists of a high plateau, covered by a permanent ice-sheet. Its surface is constantly swept by blizzards, and its rocky shores are free of snow only in a few places in summer. In spite of the severe conditions, most of the great nations have now established scientific research stations, supplied by air, and there is now 'permanent' habitation by a few hundred people (fig. 8.24).

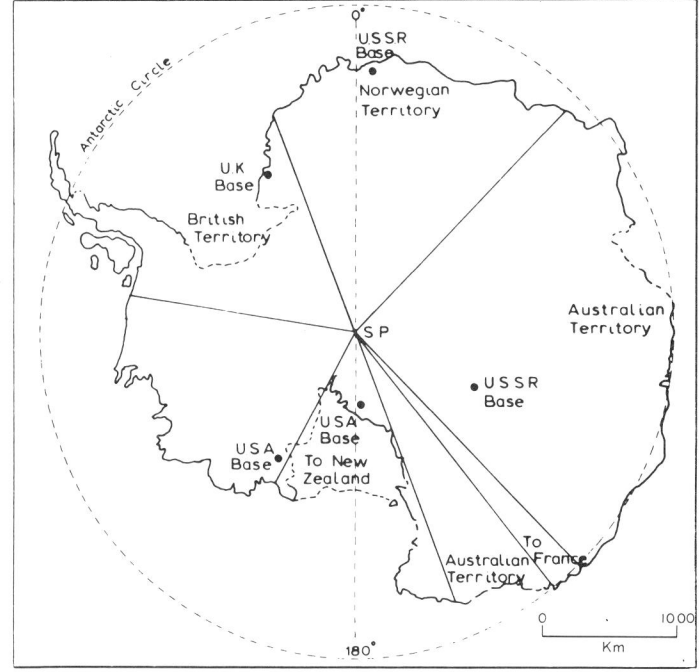

Fig. 8.24. The Antarctic.

209

Work to do

1. *Study the pictures, and answer the following questions:*

   i) *What evidence is there about weather?*
   ii) *Describe carefully the icefields.*
   iii) *Write a description of one of the pictures.*

2. *For what reasons has the government of the U.S.S.R. tried to develop the tundra?*

3. *The following are the temperature and rainfall figures for Spitzbergen (78° N 14° E).*

	Jan.	Feb.	March	Apr.	May	June
Temperature, °C	−15	−17·2	−16·2	−13·3	−8·9	2·2
Rainfall, mm	36	33	28	23	13	10

	July	Aug.	Sept.	Oct.	Nov.	Dec.
Temperature, °C	6·1	5·0	0·0	−6·1	−12·2	−13·9
Rainfall, mm	15	23	25	30	25	38

   i) *List the months with temperatures below freezing. What season do they form?*
   ii) *Which months form summer?*
   iii) *If temperatures are below freezing 'rain' is likely to fall as snow. In which months will this happen?*
   iv) *It is too cold for much precipitation in this area. What is the year's total in millimetres?*

4. *'When he had last seen the base two months before, it had been a thriving little community built round the large wooden hut which was being erected for living quarters. There had been bunches of saxifrage then and clumps of white flowers among the rocks behind the scattered tents where everyone was sleeping until the permanent living quarters were finished. Now all that he remembered about it had vanished . . . had disappeared under the mantle of snow, and the whole valley was nothing but a vast, ice-bound bowl. To the north and west, the land was white and featureless, a bare unrelieved frozen desert stretching for hundreds of kilometres.'*

   i) *This description is of part of Greenland; of what vegetation zone is the land part?*
   ii) *In what season had he last seen the area? How can you tell?*
   iii) *What season is described now? Remember what is said in the first sentence.*
   iv) *What two words sum up the tundra land in winter?*

5. *Write a paragraph explaining why it is unlikely that Arctic lands will ever support a dense population.*

Index

Limestone 66, 89, 93, 94
Linseed oil 89, 200
Little Tosson Farm 60, 61
Liverpool 39, 117, 118, 120, 131, 135, 144
Locarno 101, 102
Locks 47, 139
Locomotives 80, 135, 136
Locusts 14, 16
London 5, 38, 39, 87, 94, 97, 104, 105, 106, 109, 110, 112, 121, 124, 127, 128, 131, 135, 144, 148, 149, 165, 170, 173, 199
Longitude 140
Lorries 124, 137
Los Angeles 5, 108, 122, 129, 130, 156
Lowland vii, 3, 67, 74, 85, 182, 183, 193
Lubricants 20
Lucerne 52
Lulea 25
Lumbering 19, 20
Luton 96

Mackenzie River 208
Magadan 208
Maggia, River 55, 101
Maggiore, Lake 55, 101
Mahogany 17, 172, 173, 174
Maize 65, 67, 69, 70, 174, 179, 193, 199, 201
Malang 30
Malaria 15, 16, 174, 176
Malaya 26, 73, 77, 90, 169
Malaysia 74, 78, 175
Malnutrition 159
Manaus 174
Manchester 40, 96, 108, 116, 131, 135, 137
Manchester Ship Canal 96, 116
Manchuria 164
Mangoes 70, 72, 174
Mangrove 174
Manhattan 109, 110
Manitoba 199
Manufacturing industries 156
Manure 60, 70, 71, 77
Maple 17, 22
Marathi 169
Margarine 174
Market 61, 64, 70, 72, 104, 106, 109, 110, 124, 158, 186, 198
Market-garden 63, 109
Marseilles 116, 124, 155
Marsh 16, 43, 44, 46, 94, 108, 113, 132, 148
Marshalling yards 134, 135
Maryland 119
Mauritius 65, 164, 166
Meanders 45, 139

Meat 65, 109, 140
Mediterranean lands (see also Climate, Mediterranean) 10, 48, 67, 124, 192–6
Mediterranean Sea 66, 132
Medway 82
Meekatharra 41, 179
Megalopolis 122
Melbourne 116
Mersey, River 96
Merseyside 96
Metal working 82, 85
Metalling 124
Meteorological Office (see also Weather stations) 7
Mexico 29, 33, 83, 154, 161
Mexico, Gulf of 10, 143
Miami 1
Michigan 163
Michigan, Lake 94
Middle East 142
Midlands 87, 90, 94, 96, 118, 128
Migration 14, 162
Migratory workers 67
Mildura 52
Milford Haven 117
Milk 60, 63, 109, 179, 209
Millet 50, 51, 69, 70, 186
Milling 25
Minerals 23–9, 80, 89, 144
Minneapolis 200
Mining 24, 25, 28, 29, 82, 85, 101, 131, 156, 169, 208
Mining, open-cast 23, 24, 25
Mining, shaft 24
Mississippi River 10, 45, 89, 108, 139, 143
Missouri River 10
Mixed farm 63, 64
Mongoloid race 167
Monsoon 6, 48, 71, 139, 155, 160, 175, 177, 182–6
Monterey Bay 192
Montreal 155
Moorlands 43
Morocco 161
Morpeth 61
Moscow 97, 112, 122, 155
Mosquito 15, 16, 141, 209
Motorways 127, 129, 149
Mountain, fold 30, 32, 33
Mud-flats 43
Murmansk 207
Murray–Darling Basin 51, 52, 197, 202
Murray River 11, 52, 139
Murrumbidgee River 52

Naples 166

Narrandera 52
Narvik 25
Natal 164
Nebraska 198
Negev 154
Negroid race 167
Nepal 40
Netherlands vii, 47, 65, 83, 155, 175
Neuchatel 168
Nevada 188
New England 122, 162
New Jersey 108, 109
New Orleans 108
New South Wales 11, 51, 52, 56, 64, 163
New York 94, 107, 108, 109, 111, 121, 122, 128, 129, 141, 144, 155, 163, 200
New Zealand 10, 31, 64, 65, 74, 75, 157, 164, 165
Newcastle 61, 131, 137
Newfoundland 5
Niagara Falls 56, 57, 58, 139
Niger, River 14, 174
Nigeria 27, 71, 83, 126, 173, 174, 175, 178
Nile, River 41, 45, 50, 51, 187
Nitrates 60
Nitrogen 11
Nomads 71, 100, 152, 154, 189, 198, 207
Norfolk 90, 127
Norman Wells 208
North America 12, 91, 100, 139, 152, 155, 157, 163, 174, 175, 188
North Atlantic 146
North Atlantic Drift 206
North Circular Road 97
North Dakota 199, 200
North Pole 149, 206
North Sea 27, 87
North Vietnam 158
Northamptonshire 24
Northern Ireland 21
Northern Territory (Australia) 180
Northumberland 60
Norway vii, 22, 25, 126, 135, 203
Norwich 137
Norilsk 207
Novy Port 207
Nuclear power station 41, 54
Nullarbor Plain 132
Nursery 74, 77, 204
Nylon 80, 91

Oaks 17, 185
Oakland Bay 126
Oases 41, 48, 71, 154, 188, 189
Oats 5, 52, 60, 62, 69, 202

Ob, River 207, 208
Oceania 157
Ohio 94
Ohio River 143
Oil 23, 26, 29, 34, 49, 80, 85, 86, 87, 92, 96, 109, 117, 139, 140, 143, 144, 155, 160, 176, 189, 190, 208
Oil-refineries 82, 86
Oil-tankers 142
Oil well 26, 85
Oklahoma 9
Olive oil 66, 192, 194
Olives 66, 67, 193, 194
Onions 67, 69, 179, 193
Ontario 19, 20, 204
Open-hearth furnace 94
Oranges 6, 67, 70, 185, 192, 193, 195
Orchards 63, 64, 192
Ordnance survey 124
Ore 24, 25, 26, 27, 28, 80, 89, 94, 137
Oregon 19, 64
Orinoco, River 179
Osaka 184
Overburden 23, 24
Overgrazing 10, 11, 66, 71, 178, 185
Overpopulation 176, 186
'Overspill' 113
Owen Falls 58
Oxen (see also Cattle) 48, 51, 68, 185, 186
Oxford 106
Oxford Street 104, 109
Oxus, River 187

Pacific Ocean 140, 149, 184, 192
Packaging 20
Padi fields 67, 186
Paint 80, 95, 174, 200
Pakistan 158, 161, 182, 183, 186
Pakistan, East 114
Pakistan, West 58
Palagnedra 55
Palm Beach 1
Palm oil 70, 72, 75, 89, 174, 175
Pampas 136, 201, 202
Pan American Highway 128
Panama Canal 28, 140, 141
Paper 20, 22, 80, 97, 203, 204
Paper mills 40, 96, 204
Paper pulp 20, 22, 82, 203, 204
Paris 109, 112, 121, 124, 144, 149
Pasture 60, 62, 63, 71, 178, 189, 197, 207
Pawpaws 72
Peas 8, 69, 70, 186, 200
Peaches 66, 67, 192, 193
Peanuts 179

214

216